NANOTECHNOLOGY SCIENCEAND TECHNOLOGY

NANOMATERIAL RESEARCH STRATEGY

NANOTECHNOLOGY SCIENCE AND TECHNOLOGY

Safe Nanotechnology
Arthur J. Cornwelle
2009. ISBN: 978-1-60692-662-8

National Nanotechnology Initiative: Assessment and Recommendations
Jerrod W. Kleike (Editor)
2009. ISBN: 978-1-60692-727-4

Nanotechnology Research Collection - 2009/2010
James N. Ling (Editor)
2009. ISBN: 978-1-60741-293-9 (DVD edition)
2009. ISBN: 978-1-60741-292-2 (PDF edition)

Strategic Plan for NIOSH Nanotechnology Research and Guidance
Martin W. Lang
2009. ISBN: 978-1-60692-678-9

Safe Nanotechnology in the Workplace
Nathan I. Bialor (Editor)
2009. ISBN: 978-1-60692-679-6

Nanotechnology in the USA: Developments, Policies and Issues
Carl H. Jennings (Editor)
2009. ISBN: 978-1-60692-800-4

Nanotechnology: Environmental Health and Safety Aspects
Phillip S. Terrazas (Editor)
2009. ISBN: 978-1-60692-808-0

New Nanotechnology Developments
Armando Barrañón (Editor)
2009. ISBN: 978-1-60741-028-7

Electrospun Nanofibers and Nanotubes Research Advances
A. K. Haghi (Editor)
2009. ISBN: 978-1-60741-220-5
2009. ISBN: 978-1-60876-762-5 (E-book)

Carbon Nanotubes: A New Alternative for Electrochemical Sensors
Gustavo A. Rivas, María D. Rubianes, María L. Pedano, Nancy F. Ferreyra, Guillermina Luque and Silvia A. Miscoria
2009. ISBN: 978-1-60741-314-1

Polymer Nanocomposites: Advances in Filler Surface Modification Techniques
Vikas Mittal (Editor)
2009. ISBN: 978-1-60876-125-8

Nanostructured Materials for Electrochemical Biosensors
Yogeswaran Umasankar, S. Ashok Kuma and Shen-Ming Chen (Editors)
2009. ISBN: 978-1-60741-706-4

Magnetic Properties and Applications of Ferromagnetic Microwires with Amorphous and Nanocrystalline Structure
Arcady Zhukov and Valentina Zhukova
2009. ISBN: 978-1-60741-770-5

Electrospun Nanofibers Research: Recent Developments
A.K. Haghi (Editor)
2009. ISBN: 978-1-60741-834-4

Nanofibers: Fabrication, Performance, and Applications
W. N. Chang (Editor)
2009. ISBN: 978-1-60741-947-1
2009. ISBN: 978-1-61668-288-0 (E-book)

Bio-Inspired Nanomaterials and Nanotechnology
Yong Zhou (Editor)
2009. ISBN: 978-1-60876-105-0

Nanotechnology: Nanofabrication, Patterning and Self Assembly
Charles J. Dixon and Ollin W. Curtines (Editors)
2010. ISBN: 978-1-60692-162-3

Gold Nanoparticles: Properties, Characterization and Fabrication
P. E. Chow (Editor)
2010. ISBN: 978-1-61668-009-1
2010. ISBN: 978-1-61668-391-7 (E-book)

Micro Electro Mechanical Systems (MEMS): Technology, Fabrication Processes and Applications
Britt Ekwall and Mikkel Cronquist (Editors)
2010. ISBN: 978-1-60876-474-7

Nanomaterials: Properties, Preparation and Processes
Vinicius Cabral and Renan Silva (Editors)
2010. ISBN: 978-1-60876-627-7

Nanopowders and Nanocoatings: Production, Properties and Applications
V. F. Cotler (Editor)
2010. ISBN: 978-1-60741-940-2

Barrier Properties of Polymer Clay Nanocomposites
Vikas Mittal (Editor)
2010. ISBN: 978-1-60876-021-3

Phage Display as a Tool for Synthetic Biology
Santina Carnazza and Salvatore Guglielmino
2010. ISBN: 978-1-60876-987-2

Nanomaterials Yearbook - 2009. From Nanostructures, Nanomaterials and Nanotechnologies to Nanoindustry
Gennady E. Zaikov and Vladimir I. Kodolov (Editors)
2010. ISBN: 978-1-60876-451-8

Nanoparticles: Properties, Classification, Characterization, and Fabrication
Aiden E. Kestell and Gabriel T. DeLorey (Editors)
2010. ISBN: 978-1-61668-344-3

Nanoporous Materials: Types, Properties and Uses
Samuel B. Jenkins (Editor)
2010. ISBN: 978-1-61668-182-1
2010. ISBN: 978-1-61668-650-5 (E-book)

Silver Nanoparticles: Properties, Characterization and Applications
Audrey E. Welles (Editor)
2010. ISBN: 978-1-61668-690-1
2010. ISBN: 978-1- -61728-062-7 (E-book)

TiO2 Nanocrystals: Synthesis and Enhanced Functionality
Ji-Guang Li , Xiaodong Li and Xudong Sun
2010. ISBN: 978-1-60876-838-7

Mechanical and Dynamical Principles of Protein Nanomotors: The Key to Nano-Engineering Applications
A. R. Khataee and H. R. Khataee
2010. ISBN: 978-1-60876-734-2

Nanomaterial Research Strategy
Earl B. Purcell (Editor)
2010. ISBN: 978-1-60876-845-5

Magnetic Pulsed Compaction of Nanosized Powders
G.Sh Boltachev, K.A Nagayev, S.N. Paranin, A.V. Spirin and N.B. Volkov
2010. ISBN: 978-1-60876-856-1

Nanostructured Conducting Polymers and their Nanocomposites: Classification, Properties, Fabrication and Applications
Ufana Riaz and S.M. Ashraf
2010. ISBN: 978-1-60876-943-8

Bioencapsulation in Silica-Based Nanoporous Sol-Gel Glasses
Bouzid Menaa, Farid Menaa, Carla Aiolfi-Guimarães and Olga Sharts
2010. ISBN: 978-1-60876-989-6

Ion-Synthesis of Silver Nanoparticles and their Optical Properties*
Andrey L. Stepanov
2010. ISBN: 978-1-61668-862-2

ZnO Nanostructures Deposited by Laser Ablation
M. Martino, D. Valerini, A.P. Caricato A. Cretí, M. Lomascolo and R. Rella
2010. ISBN: 978-1-61668-034-3

Development and Application of Nanofiber Materials
Shou-Cang Shen, Wai-Kiong Ng, Pui-Shan Chow and Reginald B.H. Tan
2010. ISBN: 978-1-61668-931-5
2010. ISBN: 978-1-61668-829-5 (E-book)

Polymers as Natural Composites
Abdulakh K. Mikitaev; Georgii V. Kozlov and Gennady E. Zaikov (Editors)
2010. ISBN: 978-1-61668-168-5
2010. ISBN: 978-1-61668-886-8 (E-book)

Synthesis and Engineering of Nanostructures by Energetic Ions
Devesh Kumar Avasthi and Jean Claude Pivin (Editors)
2010. ISBN: 978-1-61668-209-5

Biocompatible Nanomaterials: Synthesis, Characterization and Applications
S. Ashok Kumar, Sea-Fue Wang and Soundappan Thiagarajan (Editors)
2010. ISBN: 978-1-61668-677-2
2010. ISBN: 978-1-61728-078-8 (E-book)

From Gold Nano-Particles Through Nano-Wire to Gold Nano-Layers
V. Švorčík, Z. Kolská, P. Slepička and V. Hnatowicz
2010. ISBN: 978-1-61668-316-0
2010. ISBN: 978-1-61668-722-9 (E-book)

Phase Mixture Models for the Properties of Nanoceramics
Willi Pabst and Eva Gregorova
2010. ISBN: 978-1-61668-673-4
2010. ISBN: 978-1-61668-898-1 (E-book)

Applications of Electrospun Nanofiber Membranes for Bio-separations
Todd J. Menkhaus, Lifeng Zhang and Hao Fong
2010. ISBN: 978-1-60876-782-3

Nanostructured Materials: Classification, Properties and Fabrication
Anees A. Ansari, M. Naziruddin Khan, M. Alhoshan, A.S. Aldwayyan and M.S. Alsalhi
2010. ISBN: 978-1-61668-763-2
2010. ISBN: 978-1-61728-474-8 (E-book)

Dynamics of Infiltration of a Nanoporous Media with a Nonwetting Liquid*
V. D. Borman and V. N. Tronin
2010. ISBN: 978-1-61668-8653

Low-K Nanoporous Interdielectrics: Materials, Thin Film Fabrications, Structures and Properties
Moonhor Ree, Jinhwan Yoon and Kyuyoung Heo
2010. ISBN: 978-1-61668-749-6
2010. ISBN: 978-1-61728-318-5 (E-book)

Ba(Ti,Zr)O3 – Functional Materials: From Nanopowders to Bulk Ceramics
Adelina Ianculescu and Liliana Mitoseriu
2010. ISBN: 978-1-61668-752-6
2010. ISBN: 978-1-61728-253-9 (E-book)

Gold Nanoparticles as an Antigen Carrier and an Adjuvant
L. A. Dykman, S. A. Staroverov, V. A. Bogatyrev and S. Yu. Shchyogolev
2010. ISBN: 978-1-61668-771-7
2010. ISBN: 978-1-61728-459-5 (E-book)

Energetics and Percolation Properties of Hydrophobic Nanoporous Media
V. N. Tronin and V. D. Borman
2010. ISBN: 978-1-61668-866-0
2010. ISBN: 978-1-61728-461-8 (E-book)

Superparamagnetic Iron Oxide Nanoparticles: Synthesis, Surface Engineering, Cytotoxicity and Biomedical Applications
Morteza Mahmoud, Pieter Stroeve and Abbas S. Milani
2010. ISBN: 978-1-61668-964-3

NANOTECHNOLOGY SCIENCEAND TECHNOLOGY

NANOMATERIAL RESEARCH STRATEGY

EARL B. PURCELL
EDITOR

Nova Science Publishers, Inc.
New York

For permission to use material from this book please contact us:
Telephone 631-231-7269; Fax 631-231-8175
Web Site: http://www.novapublishers.com

LIBRARY OF CONGRESS CATALOGING-IN-PUBLICATION DATA

Nanomaterial research strategy / editor, Earl B. Purcell.
p. cm.
Includes index.
ISBN 978-1-60876-845-5 (hardcover)
1. Nanostructured materials--Research. I. Purcell, Earl B.
TA418.9.N35N2536 2009
620.1'1--dc22

2009051583

Published by Nova Science Publishers, Inc. ✢ New York

CONTENTS

PREFACE

Research during the last two decades in science and engineering has resulted in the fabrication of atomically precise structures. Nanotechnology is generally defined as the ability to create and use materials, devices and systems with unique properties at the scale of approximately 1 to 100 nm. At this particle size, quantum mechanical effects often dominate and surface area per unit volume increases, resulting in materials that exhibit unique optical, mechanical, magnetic, conductive and sorptive properties. The use of nanotechnology in the consumer and industrial sectors is expected to increase significantly in the future. Nanotechnology offers society the promise of major benefits, but also raises questions of potential adverse effects. This new book discusses issues related to a research strategy.

Chapter 1 - Research during the last two decades in science and engineering has resulted in the fabrication of atomically precise structures. Nanotechnology is generally defined as the ability to create and use materials, devices and systems with unique properties at the scale of approximately 1 to 100 nm. At this particle size, quantum mechanical effects often dominate and surface area per unit volume increases, resulting in materials that exhibit unique optical, mechanical, magnetic, conductive and sorptive properties. The use of nanotechnology in the consumer and industrial sectors is expected to increase significantly in the future. Nanotechnology offers society the promise of major benefits, but also raises questions of potential adverse effects.

The challenge for environmental protection is to ensure that, as nanomaterials are developed and used, unintended consequences of exposures to humans and ecosystems are prevented or minimized. In addition, knowledge concerning how best to apply nanotechnology to detect, monitor, prevent, control, and cleanup pollution is needed.

The scope of this research document is strategic in that it discusses broad themes and general approaches. The purpose of this strategy is to guide the EPA's Office of Research and Development (ORD) program in nanomaterial research. The strategy builds on and is consistent with the foundation of scientific needs identified in the report by the Nanotechnology Environmental and Health Implications (NEHI) Working Group (NSTC, 2006), and on the EPA Nanotechnology White Paper (EPA, 2007). Special attention is given to EPA's role among federal agencies in addressing data needs for hazard assessment, risk assessment, and risk management relevant to the EPA mission and regulatory responsibilities. ORD will use the NRS and incorporate these research activities into its multi-year planning process.

Chapter 2 - Nanotechnology—a term encompassing nanoscale science, engineering, and technology—is focused on understanding, controlling, and exploiting the unique properties of matter that can emerge at scales of one to 100 nanometers. A key issue before Congress regarding nanotechnology is how best to protect human health, safety, and the environment as nanoscale materials and products are researched, developed, manufactured, used, and discarded. While the rapidly emerging field of nanotechnology is believed by many to offer significant economic and societal benefits, some research results have raised concerns about the potential adverse environmental, health, and safety (EHS) implications of nanoscale materials.

Some have described nanotechnology as a two-edged sword. On the one hand, some are concerned that nanoscale particles may enter and accumulate in vital organs, such as the lungs and brains, potentially causing harm or death to humans and animals, and that the diffusion of nanoscale particles in the environment might harm ecosystems. On the other hand, some believe that nanotechnology has the potential to deliver important EHS benefits such as reducing energy consumption, pollution, and greenhouse gas emissions; remediating environmental damage; curing, managing, or preventing diseases; and offering new safety-enhancing materials that are stronger, self-repairing, and able to adapt to provide protection.

Stakeholders generally agree that concerns about potential detrimental effects of nanoscale materials and devices—both real and perceived—must be addressed to protect and improve human health, safety, and the environment; enable accurate and efficient risk assessment, risk management, and cost-benefit trade-offs; foster innovation and public confidence; and ensure that society can enjoy the widespread economic and societal benefits that nanotechnology may offer.

Congressionally-mandated reviews of the National Nanotechnology Initiative (NNI) by the National Research Council and the President's Council of Advisors on Science and Technology have concluded that additional research is required to make a rigorous risk assessment of nanoscale materials.

In: Nanomaterial Research Strategy
Editors: Earl B. Purcell

ISBN: 978-1-60876-845-5

Chapter 1

DRAFT NANOMATERIAL RESEARCH STRATEGY (NRS)

United States Environmental Protection Agency

EXECUTIVE SUMMARY

Research during the last two decades in science and engineering has resulted in the fabrication of atomically precise structures. Nanotechnology is generally defined as the ability to create and use materials, devices and systems with unique properties at the scale of approximately 1 to 100 nm. At this particle size, quantum mechanical effects often dominate and surface area per unit volume increases, resulting in materials that exhibit unique optical, mechanical, magnetic, conductive and sorptive properties. The use of nanotechnology in the consumer and industrial sectors is expected to increase significantly in the future. Nanotechnology offers society the promise of major benefits, but also raises questions of potential adverse effects.

The challenge for environmental protection is to ensure that, as nanomaterials are developed and used, unintended consequences of exposures to humans and ecosystems are prevented or minimized. In addition, knowledge concerning how best to apply nanotechnology to detect, monitor, prevent, control, and cleanup pollution is needed.

The scope of this research document is strategic in that it discusses broad themes and general approaches. The purpose of this strategy is to guide the

EPA's Office of Research and Development (ORD) program in nanomaterial research. The strategy builds on and is consistent with the foundation of scientific needs identified in the report by the Nanotechnology Environmental and Health Implications (NEHI) Working Group (NSTC, 2006), and on the EPA Nanotechnology White Paper (EPA, 2007). Special attention is given to EPA's role among federal agencies in addressing data needs for hazard assessment, risk assessment, and risk management relevant to the EPA mission and regulatory responsibilities. ORD will use the NRS and incorporate these research activities into its multi-year planning process.

ORD has identified four key research themes and seven key scientific questions addressing each of the research themes where we can provide leadership for the federal government research program and support the science needs of the Agency.

- Sources, Fate, Transport, and Exposure
    - Which nanomaterials have a high potential for release from a life-cycle perspective?
    - What technologies exist, can be modified, or must be developed to detect and quantify engineered nanomaterials in environmental media and biological samples?
    - What are the major processes/properties that govern the environmental fate of engineered nanomaterials, and how are these related to physical and chemical properties of these materials?
    - What are the exposures that will result from releases of engineered nanomaterials?
- Human Health and Ecological Research to Inform Risk Assessment and Test Methods
    - What are the effects of engineered nanomaterials and their applications on human and ecological receptors, and how can these effects be best quantified and predicted?
- Risk Assessment Methods and Case Studies
    - Do Agency risk assessment approaches need to be amended to incorporate special characteristics of engineered nanomaterials?
- Preventing and Mitigating Risks
    - What technologies or practices can be applied to minimize risks of engineered nanomaterials throughout their life cycle, and how

can nanotechnologys' beneficial uses be maximized to protect the environment?

The anticipated outcomes from this research program will be focused research products to address risk assessment and management needs for nanomaterials in support of the various environmental statutes for which the EPA is responsible.

Major Acronym List	
AA	Assistant Administrator
ADME	absorption, distribution, metabolism, elimination
AML	Advanced Measurement Laboratory
BMPs	best management practices
CAA	Clean Air Act
CEA	comprehensive environmental assessment
CERCLA	Comprehensive Environmental Response Compensation and Liability Act
CR	Current Research
CREM	Council for Regulatory Environmental Modeling
CT	Committee on Technology
DAA	Deputy Assistant Administrator
DoD	Department of Defense
DOE	Department of Energy
DSSTox	Distributed Structure-Searchable Toxicity
HPG	Hypothalamic-Pituitary-Gonadal
IRIS	Integrated Risk Information System
LCA	Life-Cycle Analysis
MOA	mechanism of action
MOAs	Modes of Action
MR-CAT	Materials Research Collaborative Access Team
MYP	multi-year plan
NAS	National Academy of Science
NCCT	National Center for Computational Toxicology
NCEA	National Center for Environmental Assessment
NCER	National Center for Environmental Research
NCI	National Cancer Institute
NCL	Nanotechnology Characterization Laboratory
NEHI WG	Nanotechnology Environmental and Health Implications Working Group
NERL	National Exposure Research Laboratory
NHEERL	National Health and Environmental Effects Laboratory
NGO	Non-Governmental Organization

Major Acronym List (Continued)

NIEHS	National Institute of Environmental Health Sciences
NIOSH	National Institute for Occupational Safety and Health
NIST	National Institute of Standards and Technology
NNCO	National Nanotechnology Coordination Office
NNI	National Nanotechnology Initiative
NPDs	National Program Directors
NRC	National Research Council
NRMRL	National Risk Management Research Laboratory
NRS	Nanomaterial Research Strategy
NSET	Nanoscale Science Engineering and Technology
NSF	National Science Foundation
NSTC	National Science and Technology Council
OECD	Organization for Economic Cooperation and Development
ORD	Office of Research Development

1.0. INTRODUCTION[1]

The purpose of this Nanomaterial Research Strategy (NRS) is to guide the nanotechnology research program within the Environmental Protection Agency's (EPA's) Office of Research and Development (ORD). The strategy builds on and is consistent with the foundation of scientific needs identified in the report by the Nanotechnology Environmental and Health Implications Working Group (NEHI) (NSTC, 2006), and on the EPA Nanotechnology White Paper (EPA, 2007). Special attention is given to EPA's role among federal Agencies in addressing data needs for hazard assessment, risk assessment, and risk management relevant to the EPA mission and regulatory responsibilities. Key scientific questions of importance to the Agency are identified and a research program is described to address those questions. ORD will use the NRS to incorporate these research activities into its multi-year planning process. As a living document, it is expected that this strategy will be further refined in future years, based in part on the activities described herein and on other sources of new knowledge about nanomaterials.

The NRS contains sections introducing the human health and environmental issues associated with nanotechnology, provides background information on federal collaboration and ORD research accomplishments to date, and discusses the development of the research program. Since this emerging area of science is expanding at such a rapid pace, the NRS will be a flexible document that is reviewed and modified as new scientific information is published and as new issues arise for the EPA.

The use of nanotechnology in consumer and industrial sectors is expected to increase significantly in the future. Nanotechnology is defined as the ability to create and use materials, devices, and systems with unique properties at the scale of approximately 1 to 100 nanometers. Nanotechnology offers society Nanotechnology offers society the promise of major benefits, but also raises questions of potential adverse effects. At this particle size, quantum mechanical effects often dominate and surface area per mass is dramatically increased, resulting in materials that exhibit unique physical, chemical, electrical, optical, mechanical and magnetic properties. For example, gold is considered to be relatively inert, but depending on particle size, nanoscale gold particles become very reactive and can be green, red, or other hues. Beyond nanoscale versions of existing compounds, new structures such as the carbon-based fullerenes and nanotubes can now be created using nanotechnology.

The challenge for environmental protection is to ensure that as nanomaterials are developed and used, any unintended consequences of exposures to humans, ecosystems, and the environment are prevented or minimized. In addition, knowledge concerning how best to apply nanotechnology to detect, monitor, prevent, control, and cleanup pollution is needed. The key to such understanding is a strong body of scientific information, and the sources of such information are the numerous environmental research and development activities that are either currently underway or are pending within government agencies, academia, and the private sector. Collaboration and communication in this field will undoubtedly play a pivotal role in both how and when critical research questions are addressed.

Examples of the potential environmental benefits of nanotechnology and engineered nanomaterials include: early environmental treatment and remediation; stronger and lighter materials; and smaller, more accurate, and more sensitive sensing and monitoring devices. Additional benefits include: cost-effective development and use of renewable energy sources; development of processes with reduced material and energy requirements and minimal waste generation; early detection and treatment of diseases; and improved systems to control, prevent, and remediate pollution problems.

Revolutionary science and engineering advances applied to the existing infrastructure of consumer goods, manufacturing methods, and materials usage could also have unintended consequences on the environment. Members of the U.S. Congress, non-governmental organizations (NGOs), and others have expressed concern that, while the field of nanotechnology and the number of consumer products incorporating nanomaterials are increasing dramatically, in many cases; the safety of these materials has not been demonstrated.

EPA's mission is to protect human health and the environment. Therefore, understanding the consequences of nanomaterials and how they may impact human health and ecosystems is of critical importance to the Agency. This includes impacts associated with the manufacture, processing, use, and disposal or recycling of engineered nanomaterials. These impacts can occur as a result of exposure to and the toxicity of the materials themselves or altered materials as these materials interact with other compounds or the environment as they age. For instance, alterations in a materials' surface charge, morphology, coating stability, functionalization, agglomeration, etc. will affect its fate, transport, and exposure to humans and ecosystems. In fact, early toxicity studies have demonstrated changes in toxicity potential with changes in surface charge, particle size, state of agglomeration and coating type. In addition, exposure plays a pivotal role in the assessment of any potential harm from these materials. Exposure can occur during production and/or manufacturing processes of engineered nanomaterials, through their use, or when nanoproducts enter the waste stream and are distributed throughout the environment.

ORD's mission in support of this broader Agency effort is comprised of, but not limited to, the following actions.

- Performing research and development to identify, understand, and solve current and future environmental problems
- Providing responsive technical support to EPA's mission
- Integrating the work of ORD's science partners (other agencies, nations, private sector organizations, academia, and international organizations)
- Providing leadership in addressing emerging environmental issues and in advancing the science and technology of risk assessment and risk management

The initial emphasis of the NRS will be to evaluate and assess the extent to which nanomaterials and products impact the environment and human health. This focus is consistent with EPA's primary statutory responsibilities to protect human health and the environment and ORD's mission to address emerging environmental issues. Results from this research will directly inform future policy decisions regarding how to address possible adverse implications associated with the production, use, recycling or disposal of nanomaterials and nanoproducts (i.e., products containing nanomaterials). Initially, a smaller portion of the NRS proposed research will focus on beneficial environmental applications, such as more effective control technologies and enhanced production processes that reduce emissions and releases of conventional pollutants. As the program evolves over time, ORD will augment its efforts in this area.

2.0. BACKGROUND

2.1. US National Nanotechnology Initiative

The interest in research on the safety of nanomaterials extends beyond the EPA. The U.S. National Nanotechnology Initiative (NNI) is a federal effort established to coordinate the multiagency efforts in nanoscale science, engineering, and technology. The NNI is managed within the framework of the National Science and Technology Council (NSTC), the Cabinet- level council by which the President coordinates science, space, and technology policies across the federal government. The Nanoscale Science Engineering and Technology (NSET) Subcommittee of the NSTC coordinates planning, budgeting, program implementation and review of the NNI to ensure a balanced and comprehensive initiative. The NSET Subcommittee is composed of representatives from the 26 agencies participating in the NNI. Interagency management of the NNI occurs within the framework of the NSTC Committee on Technology (CT).

As the active interagency coordinating body, the NSET Subcommittee establishes the goals and priorities for the NNI and develops plans, including appropriate interagency activities, aimed at achieving those goals. The NSET Subcommittee promotes a balanced investment across all agencies, so as to address all of the critical elements that will support the responsible development and utilization of nanotechnology. The NSET Subcommittee exchanges information with academic, media, industry, and State and local government groups. A number of working groups have been formed under the NSET Subcommittee to improve the efficiency of its operations and focus interagency attention and activity. Current working groups are focused on environmental and health implications of nanotechnology, liaison with industries, nanomanufacturing, international and public engagement activities.

The National Nanotechnology Coordination Office (NNCO) provides technical and administrative support to the NSET Subcommittee, in the preparation of multi-agency planning, budget, and assessment documents. The NNCO also serves as the point of contact on federal nanotechnology activities for government organizations, academia, industry, professional societies, foreign organizations, and others. Finally, the NNCO develops and makes available printed and other materials concerning the NNI, and maintains the NNI website, www.nano. gov.

2.2. Environment, Health, and Safety Focus

One of the priorities of the NNI is to support research and development that leads to a detailed understanding of the environmental, health and safety impacts of nanomaterials and nanoproducts and the potential environmental impacts of the application of nanotechnology. The EPA Nanotechnology White Paper also indicates that research into the potential implications of nanomaterials is critical. The following provides additional rationale to support our initial focus on the implications of nanomaterials and nanoproducts:

- Studies of potential health risks of nanomaterials are supported by six federal agencies: the National Institute of Environmental Health Sciences (NIEHS) (including the National Toxicology Program); the National Institute for Occupational Safety and Health (NIOSH); EPA; the Department of Defense (DoD); the Department of Energy (DOE); and the National Science Foundation (NSF).
- The NSET interagency group was established to enable coordination among the member agencies, to identify and prioritize research needed to support regulatory decision-making, and to promote better communication among the federal government, industry, and researchers. The Nanotechnology Environmental and Health Implications (NEHI) working group established within NSET to focus on coordination of environmental, health and safety research. ORD is a member NSET and its various subgroups and also participates in international dialogue on environmental, health, and other societal issues.
- In September 2006, the NEHI working group of the NSET released a report, "Environmental, Health, and Safety Research Needs for

Engineered Nanoscale Materials," outlining the research needed for the federal government to understand and adequately address the potential risks of nanomaterials. Areas of particular interest to EPA in the NNI report include assessing exposure to nanomaterials, determining the behavior and impact of nanomaterials on the environment, understanding the fate, transport, and transformation of nanomaterials in biological systems; the ecological effects on the environment; the health effects of nanomaterials throughout living organisms; and development of sampling methods for relevant nanomaterials to evaluate potential effects. (http://nano.gov/NNI_EHS_research_needs .pdf)

- In 2004 EPA's Science Policy Council (SPC) created an Agency-wide workgroup to examine nanotechnology from an environmental perspective. The workgroup developed a Nanotechnology White Paper, which was issued in February, 2007 (EPA/100/B-07/001) http://www.epa. gov.OSA. The purpose of the White Paper is to both inform EPA management of the science issues and needs associated with nanotechnology and communicate nanotechnology science issues to stakeholders and the public.

The Nanotechnology White Paper provides:

- A basic description of nanotechnology
- Information on why EPA is interested in nanotechnology
- Potential environmental benefits of nanotechnology
- Risk assessment issues specific to nanotechnology
- A discussion of responsible development of nanotechnology and the EPA's statutory mandates
- An extensive review of research needs for health, ecological and environmental applications and implications of nanotechnology
- Staff recommendations for addressing science issues and research needs, including research needs within most risk assessment topic areas (e.g., human health and ecological effects research, fate and transport research)

One of the Nanotechnology White Paper appendices describes EPA's framework for nanotechnology research, which outlines the strategic focus of the research program. The goal of EPA's nanotechnology research effort is to provide key information on environmental implications and potential beneficial environmental applications to complement other federal, academic, and private-sector research activities. Appendix A of the NRS presents a side-by-side table that summarizes the research needs from the EPA White Paper and a corresponding column that lists ORD current research (CR), short-term research (SR) and long-term research (LR) activities. In addition, the Agency is actively engaged in the pursuit of knowledge by:

- Supporting in-house and extramural research
- Organizing scientific workshops, symposia and conferences
- Coordinating with stakeholders—including industry, academia, NGOs, other federal agencies, and international organizations—to obtain information and enhance coordination and collaboration
- Coordinating within EPA to ensure that the right questions are asked and the right data are obtained that will address the various statutory mandates for environmental protection

In addition to the NEHI research needs document and the EPA Nanotechnology White Paper, a number of national and international stakeholders have published articles and reports that highlight the need for research related to the environmental, health, and safety aspects of nanotechnology (Maynard, 2006). There is clearly global interest in understanding and managing the risks of this emerging technology so that its many potential benefits, including those for the environment, may be realized.

2.3. EPA Regulatory Role

Regulatory decision making in EPA requires risk managers to have sufficient information on risk and the social and economic implications of various control options before making decisions. Informing the risk manager of risk and options for controlling risk so that wise decisions can be made has been further codified in the Presidential/Congressional Commission on Risk Assessment and Risk Management (Presidential/Congressional Commission, 1997). This Commission developed a general framework for risk and risk management designed to work in a variety of situations, but primarily intended for risk decisions related to setting standards, controlling pollution, protecting health, and cleaning up the environment. The framework shown in Figure 2-1 puts health and environmental problems in their larger, real-world context. In this framework, the process begins by defining the problem. Then the risks associated with the problem are analyzed followed by an examination of the options for addressing the risks. Decisions are made about which option to implement and actions to take to implement the decisions. Measurement techniques are developed to allow determination of the extent of the problem—both prior to, and after, the actions. Finally, an evaluation of the action's results is conducted. All of this is carried out in collaboration with stakeholders at every step possible.

Regulatory decisions regarding nanomaterials are covered under current statutes. EPA intends to review nanomaterial products and processes, pursuant to its authorities under the Toxic Substances Control Act (TSCA), the Federal Insecticide, Fungicide and Rodenticide Act (FIFRA), the Clean Air and Water Acts (CAA and CWA), the Safe Drinking Water Act (SDWA) Comprehensive Environmental Response Compensation and Liability Act (CERCLA) and Resource Conservation and Recovery Act (RCRA). Under the Toxic Substances Control Act (TSCA), Premanufacture Notices must be submitted to the EPA by an entity that wishes to manufacture or import new chemical substances that are not currently on the TSCA Inventory of Chemical Substances. There is some question as to whether nanomaterials are "new" compounds. Under FIFRA nanomaterials added to an existing pesticide product may require reapproval, and the EPA must determine whether the altered product might cause unreasonable adverse effects on the environment including human health risks. The CAA allows for the development of air quality criteria for pollutants anticipated to endanger public health and welfare, mandates the identification of the sources and the issuance of technology-based emissions standard for 189 pollutants, and requires that any mobile source fuel or additive be registered. Risks from airborne nanomaterials may reasonably need assessing in all of these areas. Wastewater streams containing nanomaterials might be controlled through effluent limits in permits established under the CWA. If nanomaterials enter drinking water they may be subject to regulation using Maximum Contaminant Level Goals and Maximum Contaminant Levels under SDWA. Risks from nanomaterials in waste sites would be evaluated and controlled under the authority of CERCLA and RCRA. Figure 2-2 highlights the information needs of the major statutes that EPA administers.

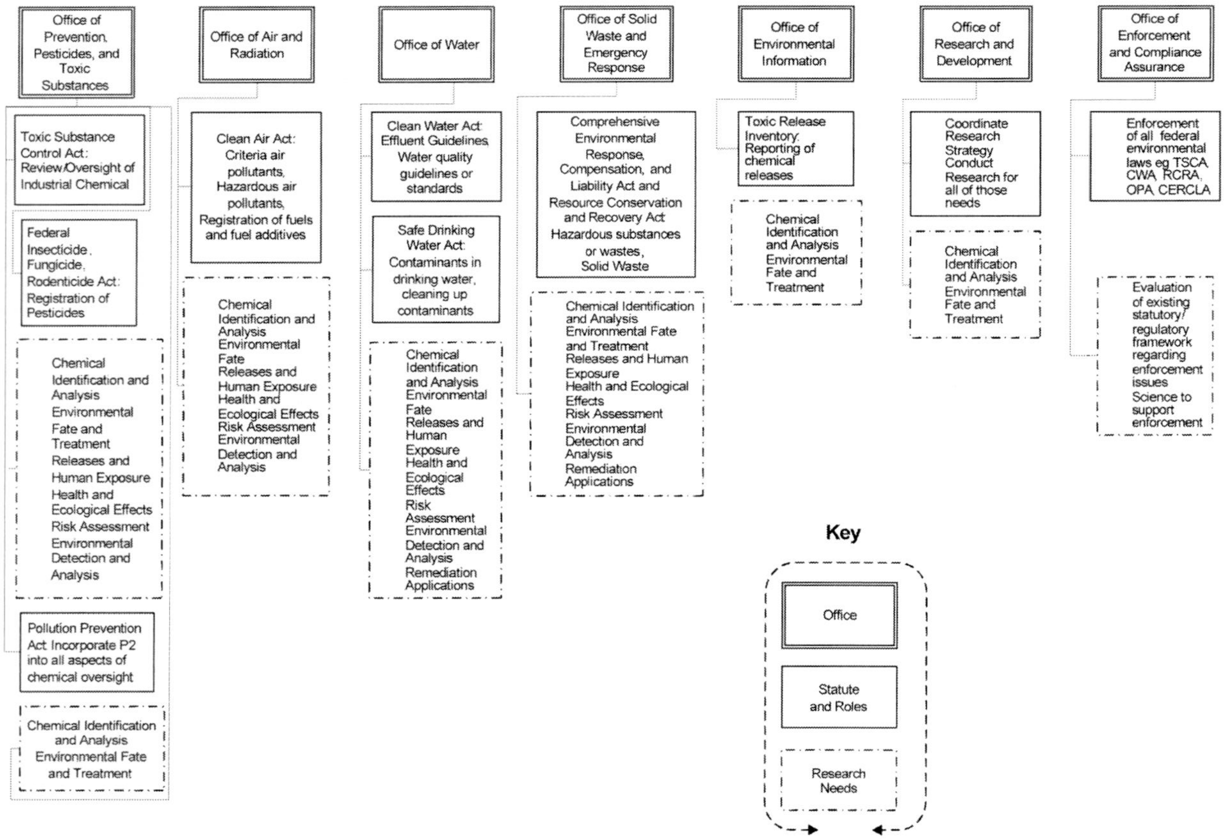

Figure 2-2. EPA Office Roles, Statutory Authorities, and Categories of Research Needs Related to Nanotechnology.

Figure 2-1. Presidential/Congressional Commission on Risk Assessment and Risk Management's Framework for Environmental Health Risk Management.

2.4. ORB Research Accomplishments

2.4.1. ORB Science to Achieve Results (STAR) Program

The extramural research program at EPA has taken a holistic approach to studying nanotechnology, targeting research toward the identification of the beneficial applications of nanotechnology and seeking to increase data and understanding of potential effects. Science To Achieve Results (STAR) grants and Small Business Innovation Research (SBIR) contracts were designed to generate exposure, fate/transport, and human and eco-toxicity data, pursue novel pollution prevention and environmentally benign manufacturing and processing techniques, and assist in the development of novel treatment and remediation technologies.

The objective of STAR is to meet the data needs of the various EPA offices as well as those of other agencies, the scientific community, and the general public. This approach allows the Agency to play an important role in supporting the development of new technologies that could improve the environment, as well as in ensuring that new materials and compounds do not pose unreasonable risks to humans or the environment. One of the ways that the Agency supports research is through its STAR competitive grants program, managed by the National Center for Environmental Research (NCER) in ORD. The objective of STAR is to meet the data needs of the various EPA program offices, as well as those of other agencies, the scientific community, and the general public. Grants funded through the STAR program have focused on both the applications and implications of nanotechnology use.

The initial grants funded by STAR in 2002 were primarily on applications. Since 2002, ORD has funded 35 grants focused on using nanotechnology to address environmental challenges. The areas of research include green manufacturing, contamination remediation, sensors for environmental pollutants, and waste treatment. Focus has shifted to implications as interest in gathering data on the safety of nanomaterials have grown. By 2008, the STAR program had funded more than $29 million for 86 research projects on the environmental applications and implications of nanotechnology.

Since 2004 STAR grants have been issued in collaboration with other federal agencies including NSF, NIEHS, NIOSH, and DOE. ORD works with these agencies to identify and issue calls for proposals in areas related to human and environmental health. EPA expects that future calls for proposals will involve other federal agencies, as well as international organizations, such as the European Commission. Future STAR research calls for proposals that will seek to facilitate collaborations between ORD researchers and STAR researchers. This will result in strengthening extramural research through the expertise of Agency scientists and will also strengthen the Agency in-house nanotechnology research efforts, which are in their initial stages.

Figure 2-3 shows STAR research funded to date in nanotechnology. Although it is apparent from this figure that the bulk of the STAR research (up until the publication date of this research strategy) falls within the categories of fate/transport and toxicity, the Agency's nanotechnology research strategy will focus on fate/transport and exposure. Furthermore, this research will be conducted from a complete life cycle perspective to facilitate research and improve understanding of the effects of nanomaterials, enabling appropriate risk assessment and management strategies to be developed. These are focus areas where the Agency can be most effective and have the most impact while also playing a key role in the remaining areas by coordinating with other federal agencies.

Table 2.1. STAR Grants for Nanotechnology Applications

Research category	Number of grants	Award totals
Green manufacturing	7	$2,393,000.00
Remediation	10	$3,433,394.00
Sensors	13	$4,564,000.00
Treatment	5	$1,817,089.00
TOTAL	**35**	**$12,207,483.00**

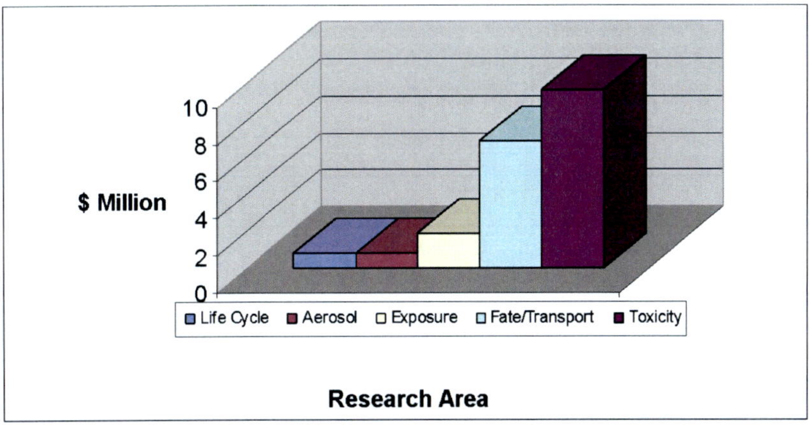

Figure 2-3. STAR Grant Research Funding Areas

The number of grants related to the implications of nanotechnology use that have been funded by the STAR program has increased significantly over the years. In 2002 there were two grants funded for air research only, and in 2003 two were funded for life-cycle analysis (LCA). In 2004, 2005, and 2006 EPA funded 12, 14, and 21 grants, respectively. The Agency will continue to concentrate extramural support on implications research in the near future. EPA expects that future calls for proposals will be done in collaboration with other federal agencies, as well as international agencies, such as the European Commission. The table below categorizes the nano implications studies into those that address the potential human health and environmental effects, respectively. Each “x” in the table indicates that the study includes a particular endpoint and material class as described in the research protocol. A single study could be represented more than once in the table, and a single "x" may represent a number of compounds in a particular materials class. For example, if a study were to include several metal oxides in the research protocol, all of these compounds would be represented by one "x."

In addition to the STAR funded nanotechnology research grant program, EPA supports nanotechnology research conducted by small businesses. EPA's Small Business Innovation Research (SBIR) program has funded 49 projects for over $5 million in funding related to nanotechnology development, nanomaterials, and clean technology. These projects range from a nanocomposite-based filter for arsenic removal in drinking water to nanofibrous manganese dioxide for emission control of volatile organic compounds (VOCs). The SBIR program is also interested in technologies that utilize nanotechnology to detect conventional pollutants in aqueous, air, and soil environments.

For a full list of nanotechnology projects funded under EPA's SBIR program, please visit: http://es.epa.gov/ncer/nano/research/sbir_index.html

Study Focus	**Material Class**			
	Carbon Nanotubes	**Fullerenes**	**Metal-Based**	**Other***
Cytotoxicity	XXXX	XX	XXXX	XXX
General toxicity				
Dermal	X	XX	XX	
Pulmonary	XXXXX	X	XXXX	
Translocation/Disposition	X	X	XXX	

2.4.2. ORD's In-house Research Program

Within the in-house research program, ORD has to date engaged in limited research related to nanotechnology including:

- Developing low-emitting coating formulations using nanopolymers: ORD researchers developed a novel family of modified "hyperbranched" polymers, which were successfully formulated with commercial resin for auto refinishing.
- Evaluating pollution prevention potential: ORD researchers have evaluated the potential to reduce or eliminate waste from

manufacturing processes and foster better materials; allowing more efficient production, effective break down of hazardous material, and providing alternatives to solvents or high temperature processes that can damage the environment.

- Testing ofnanotechnology membranes: ORD researchers used the Fyne Process nanofiltration system (PCI Membrane Systems Inc.), to reduce total organic carbon (TOC) concentration in source water by more than 95%.
- Studying iron-based permeable reactive barriers: ORD research in this area involves in-situ remediation of contaminant plumes in ground-water systems. This work improved the understanding of the dynamics of iron corrosion relative to the rates and sustainability of beneficial reactions that effect contaminant removal or transformation.
- Researching health effects: Using various cellular models, ORD researchers have examined the *in vitro* pulmonary toxicity of carbon nanotubes as well as the neurological toxicity of nano TiO2. These studies have shown: 1) unique gene expression patterns within airway cells exposed to carbon nanotubes versus environmental particles; 2) surface modifications influenced carbon nanotube *in vitro* pulmonary toxicity; and 3) cellular oxidative stress to be a mechanism of nano TiO2 induced toxicity in brain microglia cells.

2.5. Collaboration/ Leveraging

EPA is leveraging its research and development efforts by partnering with other federal agencies such as NIH/NIEHS, NIOSH, NSF, and DOE, which are also conducting or supporting research on the toxicity and human health effects of nanomaterials. The Agency is also coordinating with the National Institute of Standards and Technology (NIST) and the National Institutes of Health/Nanotechnology Characterization Laboratory (NIH/NCL), which are conducting key research on nanomaterial metrology, characterization, and detection devices. Seeking to continually broaden its collaborative efforts, EPA coordinates its research activities with the Nanotechnology Environmental and Health Implications working group of the Nanoscale Science, Engineering, and Technology subcommittee of the NSTC. Figure 2-4 illustrates the various federal sources of scientific information for use in EPA decisions.

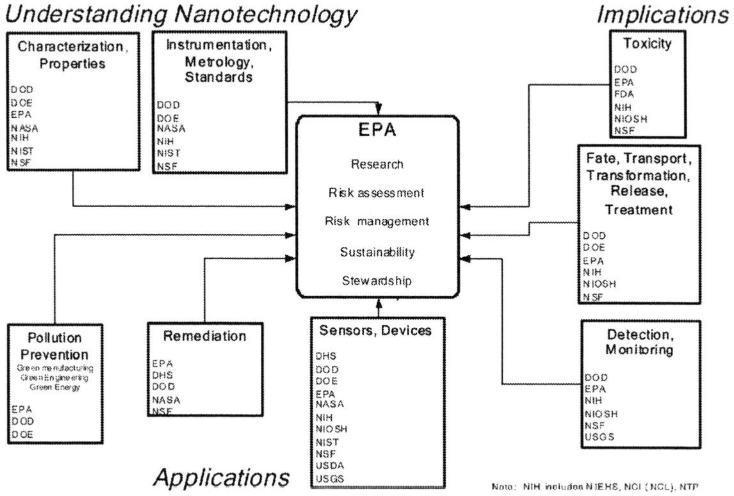

Figure 2-4. Federal Sources to Inform EPA's Nanotechnology Activities.
(Based on information in the NNI Supplement to the 2006 and 2007 budget and other information.).

Internationally, EPA plays a leading role in the nanomaterial testing/test guidelines efforts of the Organization for Economic Cooperation and Development (OECD). Future extramural research efforts involve collaborating with the European Commission, Japan, China, Singapore, and Taiwan, among others. Each of these entities has an important role to play in meeting the considerable global research needs related to nanotechnology and the environment. Because EPA's research and development program is focused on risk assessment and management to support Agency decisions, as well as on research areas not addressed by other agencies, the impact on the total federal research enterprise of the scientific information generated by our laboratories and centers is disproportionately high relative to the size of our program. This will enable the Agency to continue to provide strong leadership in the area of nanotechnology EHS both nationally and internationally.

3.0. Research Strategy Overview

The purpose of ORD's research program in support of the National Nanotechnology Initiative is to conduct focused research to address risk assessment and risk management needs for nanomaterials in support of the various environmental statutes for which the EPA is responsible. This program will be coordinated with research conducted by other federal agencies, where the EPA will lead selected research areas and rely on research products under the leadership of its federal research partners in other research areas. Collaboration is encouraged among researchers across the government, industry, and the international community.

ORD is uniquely positioned within the federal government to support the overall NNI objectives while also supporting EPA's strategic goals.

- ORD's research laboratories and centers have the expertise to integrate human health and ecological data to provide the Agency's program and regional offices with scientific information most appropriate for risk assessment and decision support.
- ORD has extensive facilities to test nanomaterials in aquatic and terrestrial ecosystems, as well as to measure and model the fate, transport, transformation, and effects of nanomaterials in environmental media.
- ORD has unique and extensive historical laboratory expertise and capacity to identify approaches to prevent and manage risks from environmental exposures to nanomaterials, including the development and verification of technologies to detect, measure, and remove nanomaterials from environmental media.
- ORD has the capability to leverage results from EPA STAR grant research, as well as collaborating with grantees to address the many challenging research issues.

3.1. ORB Scientific Expertise Applied to Nanomaterials[2]

3.1.1. Fate and Transport Expertise and Capabilities

ORD researchers have extensive expertise and experience in understanding and modeling the fate, transport, retention, and release of chemicals in various environmental media. Scientific knowledge concerning the fate and transport of compounds over a range of conditions in air, aquatic systems, soils, and sediment has been a particular strength of ORD science.

Examples of expertise in this area include:

- Residual soil contaminants at CERCLA waste sites have been evaluated by ORD scientists to determine if the contaminants would eventually migrate to the underlying aquifer.
- Assessments have been made by ORD researchers, based on considerations of the phases in which a compound is likely to occur in the atmosphere, of the degree to which the substance will respond to a group of factors that influence its fate in atmospheric, surface, or subsurface environments.
- ORD researchers have estimated the approximate lifetime in the atmosphere, soil, or water of a variety of compounds.
- ORD researchers have made accurate determinations concerning whether emitted or released compounds can be detected in the environment.
- The ways and processes by which compounds are altered as they contact other compounds, as they age, and as they enter and exit various environmental media have been extensively studied by ORD researchers.

3.1.2. Human and Ecological Effects Expertise and Capabilities

ORD's health and ecological risk assessment research within the Air, Water, and Safe Products/Safe Pesticides programs has established unique multi-disciplinary facilities and expertise that are directly applicable to addressing the health and ecological implications of nanomaterials, and their applications, resulting from various potential routes of exposure. Facilities and scientific expertise established in ORD have been called upon by Congress several times to assess the health effects of other types of particles including particulate matter in ambient air. These substantial health and ecological risk assessment-based research activities have provided critical information for Agency and regional regulatory decisions and guidance such as:

- ORD Particulate Matter (PM) (EPA/600/P-99/002af-bf) research program has significantly contributed to the Air Quality Criteria Document for PM by elucidating health and ecological effects of ultrafine, fine, and coarse ambient air, as well as source specific primary combustion particle dosimetry, fate, pulmonary/ extra pulmonary effects, hazard identification, and susceptibility factors. This research supports the CAA in setting ambient PM levels. In addition, research on the health and ecological effects of Orimulsion® (EPA/600/R-01-056a, 2001) has provided the Agency with risk assessment information on the use of alternative fossil fuels.

- Research examining the health and ecological effects of pesticides, toxic substances, as well as water borne pollutants such as arsenic has supported the FIFRA, TSCA, and the CWA.

3.1.3. Computational Toxicology Expertise and Capabilities

The EPA program in computational toxicology applies mathematical and computer models and molecular biological and chemical approaches to explore both qualitative and quantitative relationships between sources of environmental pollutant exposure and adverse health outcomes (http://www.epa.gov/comptox/index.html). This integration of modern computing with molecular biology and chemistry is allowing scientists to better prioritize data, inform decision makers on chemical risk assessments, and understand a chemical's progression from the environment to the target tissue within an organism, and ultimately to the key steps that trigger an adverse health effect. Unique capabilities that are currently available and under development through ORD's National Center for Computational Toxicology include DSSTox and ToxCast™.

The Distributed Structure-Searchable Toxicity (DSSTox) database network is creating a chemical data foundation for improved structure-activity and predictive toxicology capabilities across and outside of EPA (http://www.epa.gov/ncct/dsstox/).

The ToxCast™ program for prioritizing toxicity testing of environmental chemicals (http://www.epa.gov/comptox/toxcast/), is a new research effort in EPA to develop the ability to forecast toxicity based on bioactivity profiling and, ultimately, to develop methods of prioritizing chemicals for further screening and testing to assist EPA in the management and regulation of environmental contaminants.

3.1.4. Risk Assessment Expertise and Capabilities

ORD has a risk assessment center that focuses on implementation of the risk assessment paradigm as described by the National Academy of Sciences in its 1983 document, *Risk Assessment in the Federal Government.* This is done by providing qualitative and quantitative health hazard assessment of priority environmental contaminants for incorporation into applied risk assessments, exemplified by the Integrated Risk Information System (IRIS) toxicological reviews and summaries on reference doses, reference concentrations, oral cancer slope factors, and cancer inhalation unit risks. ORD also prepares Integrated Science Assessments (formerly Air Quality Criteria Documents) for the six criteria air pollutants. ORD has also used this capacity to respond to urgent agency priorities such as Hurricane Katrina. In addition, developing models, methods, and guidance to incorporate the latest scientific advances into EPA risk assessment practice is a continuing function of ORD. ORD' s National Center for Environmental Assessment identifies, evaluates, and conveys to the scientific community key uncertainties and research needed to improve health risk assessments through laboratory, field, and methods research. ORD also provides program office support and consultation for assessments related to air, water, waste, and pesticides. The application of risk assessment methods to nanomaterials is within the scope of ORD's past performance and current capacity.

3.1.5. Source Characterization and Risk Management Expertise and Capabilities

ORD has extensive state-of-the-art facilities and equipment as well as significant expertise that can be applied to characterize and manage releases of engineered nanomaterials. ORD's nationally and internationally recognized scientists and engineers have developed the following core engineering competencies/capabilities that can be applied to the nanotechnology issue: 1) characterization of emissions to air and releases to water and land and their subsequent movement through various media; 2) evaluation of devices and procedures to detect and characterize nanomaterials in environmental media, including identifying optimal operating conditions; 3) characterization of the effectiveness of abatement technology for emissions or effluent control; 4) identification and characterization of options to prevent pollution, including greener synthesis and manufacturing; 5) assessment of the life cycle implications of industrial and commercial products and processes; 6) verification of commercial-ready measurement and management technology; and 7) modeling efforts to evaluate the effectiveness of potential risk management options.

This work is carried out in numerous facilities across various ORD sites. These facilities can be readily deployed to address key nanotechnology science questions and perform needed research. In the air area, combustion research facilities have been used to develop, characterize, and optimize sorbents, catalysts, and other environmentally beneficial materials. These facilities have also been used to understand the fundamental mechanisms of pollutant capture and to determine the molecular scale structure-property relationships of these environmental materials. In the water and land areas, ORD has multiple multi-purpose, high- bay research facilities in Cincinnati, Ohio, which have the capacity to test and evaluate pilot and bench-scale water, wastewater, and hazardous waste treatment technologies. These facilities also have the capacity to evaluate the fate of nanomaterials in both anaerobic and aerobic landfills. The groundwater research facility in Ada, Oklahoma has the capability to evaluate nanomaterials in the subsurface soils and waters. Additionally, these facilities are a RCRA permitted treatment, storage, and disposal facility. ORD also has access to other state-of-the-art facilities and equipment through agreements with other organizations. For example, ORD researchers are participating in the creation of a new laboratory facility at Argonne National Laboratory dedicated to environmental research at the nanoscale. This facility will provide a location for cutting-edge research using X-ray spectroscopy for studies on the characterization, speciation, and behavior of inorganic contaminants at the atomic scale. It will also have the capability to examine engineered nanomaterials and assess their physical properties (e.g. structure, bonding, and surface characteristics).

3.1.6. Exposure Expertise and Capabilities

In support of the NNI, the Agency will take the lead role to assess the environmental fate and transport of nanomaterials through air, aquatic, and terrestrial ecosystems. ORD has both the expertise and capability to play a significant role in this effort. ORD has a laboratory dedicated to conducting human health and ecological exposure research that provides the tools for EPA to conduct its mission. ORD also has the capability to provide cutting edge research that addresses the most critical exposure uncertainties associated with EPA's policy decisions and to provide international scientific leadership in the area of exposure research. Exposure research is used to develop the methods, data, and models that describe our understanding of those exposures that may lead to human and ecological health risks. ORD is improving environmental quality through excellence in ecosystems and human exposure research by discovering fundamental process knowledge (research) and integrating it into state-of-the-science computational technologies and modeling systems (primarily development).

3.2. Strategic Direction of Research Themes and Science Questions

EPA is developing a nanotechnology research strategy for fiscal years 200 8-2012 that is problem-driven and focused on addressing the Agency's needs. In developing this research framework, ORD went through a prioritization process where it evaluated research recommendations from the EPA Nanotechnology White Paper and the Nanotechnology Environmental and Health Implications Interagency Working Group of the Nanoscale Science, Engineering and Technology subcommittee on Nanotechnology (NNI, 2006). ORD scientists prioritized research topics using several defining questions:

- What research themes are important to support Agency risk assessment and management activities?
- Where can ORD expertise be applied to address and lead Federal government research areas?
- How can partnerships with Federal, academic, and industry researchers enhance research activities?
- What are the key scientific questions within each research theme that need to be addressed?

ORD has identified four key research themes where it can provide leadership for the National Nanotechnology Initiative and support the science needs of the Agency.

ORD Research Themes:

- Sources, Fate, Transport, and Exposure
- Human Health and Ecological Research to Inform Risk Assessment and Test Methods
- Risk Assessment Methods and Case Studies
- Preventing and Mitigating Risks

The current priority of these research themes follows the general directions described below.

- Sources, Fate, Transport, and Exposure will be high priority from FY07 — FY10 and moderate priority in FY — FY12.

- Human Health and Ecological Research to Inform Risk Assessment and Test Methods will be a moderate priority from FY07 — FY09 and a high priority in FY10- FY12.
- Risk Assessment Methods and Case Studies will be a moderate priority in FY07 — FY08, a high priority in FY09 — FYI 1, and a moderate priority in FY12.
- Preventing and Mitigating Risks will be a moderate priority in FY07 — FY10 and a high priority in FY11 and FY12.

The following section defines the research themes and the associated key science questions.

Research Theme: Sources, Fate, Transport, and Exposure

This research theme will focus on identifying potential sources of nanomaterials in the environment, on understanding the fate and transport in environmental media, and on characterizing exposure pathways. Activities under this research theme will address research needs identified in the NEHI document (2006) and the EPA Nanotechnology White Paper.

The primary objective of research conducted under this theme will be to determine the release points of engineered nanomaterials into the environment and the physical and chemical properties controlling the transport and transformation of nanomaterials in environmental media. This work will provide the basis for prioritizing potential human health and ecological exposure pathways that warrant further exploration.

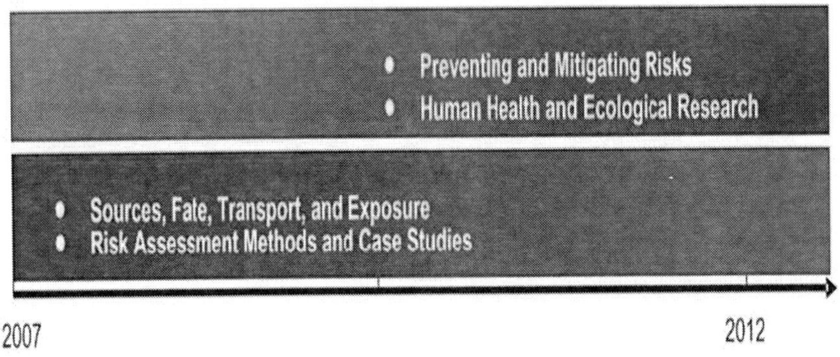

Figure 3-1. Relative Priority of Research Themes.

Key Science Questions:

1. Which nanomaterials have a high potential for release from a life-cycle perspective?
   - high-potential source characterization in industries/processes
   - identification/characterization of potentially released materials
   - characteristics and probability of byproducts
   - entry point into the environment
   - intentional "releases" such as cleanup or detection technology
   - potential release during disposal/recycling
2. What technologies exist, can be modified, or must be developed to detect and quantify engineered nanomaterials in environmental media and biological samples?
   - adequacy of existing methods/technology
   - new detection/quantification methods
   - applications of nanomaterials in new analytical/monitoring techniques
   - tools for personal or environmental monitoring
   - performance evaluation/standardization
3. What are the major processes/properties that govern the environmental fate of engineered nanomaterials, and how are these related to physical and chemical properties of those materials?
   - fate processes in air, water, soil, and biota
   - environmental modification of released materials
   - partitioning behavior
   - chemical interactions
   - environmental media interactions
   - predictive environmental models
4. What are the exposures that will result from releases of engineered nanomaterials?
   - adequacy of current exposure assessment approaches
   - exposure for human subpopulations specific eco receptors
   - early identification of potential biomarkers
   - longer term issues
   - pathways for humans and ecological receptors
   - specific routes of uptake
   - frequency, duration, and magnitude as they relate to dose parameters

Research Theme: Human Health and Ecological Research to Inform Risk Assessment and Test Methods

The diversity of nanomaterials and their applications and their ability to translocate from their initial site of deposition, represent significant challenges in assessing their human health and ecological effects. Test methods that can determine the toxicity and hazardous physical and chemical properties of nanomaterials in a validated, timely, and economic manner need to be developed. ORD 's human health and computational toxicology research programs will contribute to the development of *in vitro* test methods predictive of *in vivo* toxicity, quantitative structure-activity relationships, and other predictive models.

Similarly, for ecological testing, the EPA Nanotechnology White Paper points out that because nanomaterials are often engineered to have very specific properties, it seems reasonable to presume that they may end up having unusual toxicological effects. A number of existing test procedures that assess long-term survival, growth, development, and reproductive endpoints (both whole organism and physiological or biochemical) need to be validated for their applicability to the testing of nanomaterials.

Evaluating the adequacy of existing test methods and the development of potential new test methods to assess the toxicity of nanomaterials will complement the OECD's harmonized international test guideline efforts.

Key Science Question:

5. What are the effects of engineered nanomaterials and their applications on human and ecological receptors, and how can these effects be best quantified and predicted?
    - evaluate current test methods to assess their adequacy to determine the toxicity of nanomaterials and develop new toxicity test methods, as required
    - determine the health and ecological effects of nanomaterials including acute and chronic effects and local and systematic effects
    - determine the health and ecological effects associated with nanomaterials applications and/or interactions with environmental media, ecosystems, or other stressors

- determine if toxicity, mode(s) of action, and mechanism(s) of injury are unique to the novel physical and chemical properties of nanomaterials

- identify factors and properties regulating deposition, uptake, fate, and toxicity of nanomaterials (including hazard identification; dose-response correlations; ADME; and susceptibility/sensitivity host factors)

- identify ecological systems that have especially susceptible organisms, life stages, or populations

- develop alternative approaches/technologies/models to screen, rank, and predict the *in vivo* toxicity of nanomaterials and their applications

Research Theme: Risk Assessment

Research conducted under this theme will focus on identifying and developing risk assessment methodologies for use by Agency risk assessors that address the unique aspects of engineered nanomaterials. The EPA Nanotechnology White Paper cited a number of authors who have reviewed characterization, fate, and toxicological information for nanomaterials and proposed research for risk evaluation of nanomaterials. These publications are expected be important in developing nanomaterial risk assessment procedures.

Key Science Question:

6. Do Agency risk assessment approaches need to be amended to incorporate special characteristics of engineered nanomaterials?
    - use case studies to inform the process and refine the current strategy
    - integration of the other research areas
    - focus on how "nanoness" affects risk assessment/regulatory programs

Research Theme: Preventing and Mitigating Risks

This research theme will focus on identifying technologies or practices that can be applied to minimize exposure to engineered nanomaterials throughout their life cycle, and to investigate how nanotechnology can be applied to prevent, control, and remediate pollution. This includes studying the potential of conventional technologies to capture nanomaterials or subsequent degradation, products, materials modification to support green manufacturing of engineered nanomaterials, waste and by-product minimization, and application of nanomaterials to reduce existing environmental risks.

Key Science Question:

7. What technologies or practices can be applied to minimize risks of engineered nanomaterials throughout their life cycle, and how can nanotechnologys' beneficial uses be maximized to protect the environment?
   - materials modification
   - recycle/reuse
   - waste/byproduct minimization
   - application of nanomaterials to reduce other risks

Figure 3-2 illustrates the interrelationship of the research activities with research products informing risk assessment and management issues. Initially, research activities on the left side of the diagram will be emphasized. Because little is currently understood about the potential implications of engineered nanomaterials and products containing these materials, a life cycle perspective is proposed.

The risks associated with exposure to nanomaterials arise not only from simple ambient air or drinking water exposures. As with other production materials, engineered nanomaterials have a life cycle that includes feedstocks[3], the processing of feedstocks into manufactured nanomaterials, the distribution of nanoproducts, the storage of those products, the use of those products by consumers, and finally the recycle or disposal of the nanomaterials and waste by-products. This is commonly known as the product life cycle framework and must be considered when determining risks for nanomaterials. As shown in Figure 3-1, the consideration of the product life cycle when doing risk assessment is part of a comprehensive environmental assessment (CEA) (Davis and Thomas, 2006; Davis, 2007).

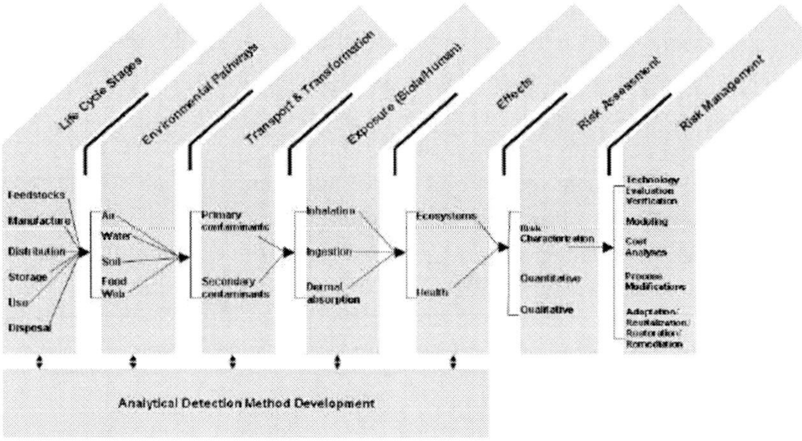

Figure 3-2. Relationship of Key Science Questions to Support Risk Assessment and Management Decisions; Based on Comprehensive Environmental Assessment (Davis and Thomas, 2006)

4.0. Research Themes

This section discusses the research themes and associated key science questions. For each science question, text addresses the topic background and program relevance, describes the proposed research activities, and discusses the anticipated outcomes.

4.1. Research Theme: Sources, Fate, Transport, and Exposure

4.1.1. Key Science Question 1: Which nanomaterials have a high potential for release from a life-cycle perspective?

4.1.1.1. Background/Program Relevance

Because they are so small, nanomaterials may be readily transported through the air, water and soil, perhaps over much greater distances than conventional materials. Uncontrolled release of these materials can occur during production, through spills, casual disposal, recycling, wastewater, agricultural operations, or weathering (of paints containing nanomaterials, for example) which may eventually lead to the presence of a large variety of nanomaterials in the environment. It may be difficult, economically unfeasible, or even impossible to remove nanomaterials from some media (e.g., surface waters or drinking water), potentially resulting in exposures to large segments of the population to complex mixtures of these materials. In order to understand the implications of nanomaterials and to identify potential approaches to manage emissions/releases, it is critical to understand potential entry points of nanomaterials into the environment. Under this question, ORD will conduct research to understand emissions/releases that can occur either during production, use, recycling, or disposal of nanomaterials. The transformation and transport of such materials once they reach the environment is addressed under Key Scientific Question 3.

Examples of points of entry into the environment include:

- **Manufacturing Waste Streams:** During the manufacture of nanomaterials, the inevitable by product and waste streams will need to be evaluated. Pollution prevention (e.g. green chemistry) research may be very helpful in the development of environmentally friendly manufacturing processes for nanomaterials.

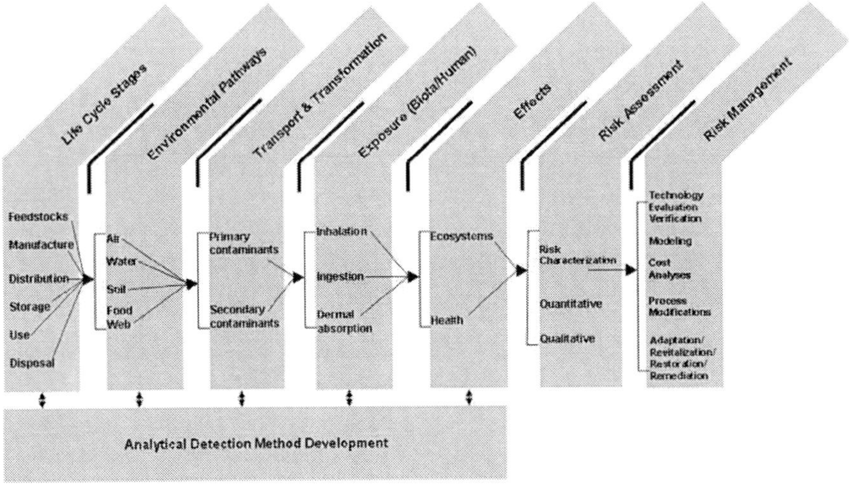

Figure 4-1. Relationship of Key Science Questions to Support Risk Assessment and Management Decisions – Life Cycle Stages.

- **Air Treatment:** Engineered nanomaterials can be emitted along with other conventional pollutants during production processes. In addition, there are products that use engineered nanomaterials where during their use nanomaterials can be emitted to the air, e.g., brakes, fuel additives.

- **Water Treatment:** Some nanomatcrials arc intended to be biocides and may disrupt drinking water treatment facilities. Personal care products and pharmaceuticals containing nanomaterials will eventually be washed down the drain and transported to wastewater treatment plants. There they will either be removed from the wastewater and end up in the biosolids residuals or they will remain in the wastewater and be discharged into surface water as part of the treatment plant's effluent.

- **Disposal of Used Material:** At the end of its useful life, each of the consumer products and equipment items created using nanomaterials will enter the waste stream. It is critical to understand where these products end up (e.g., landfill, incinerator) in order to provide guidance on possible emissions/releases of nanomaterials.

- **Product Usage:** As products incorporating engineered nanomaterials enter the consumer market place, material release may occur during the normal intended usage or conversely during unintended usage. Releases may occur through abrasion, adsorption/absorption, or volatilization, among other processes. For instance, if veterinary pharmaceuticals are administered using nanomaterials, these materials may be excreted and released into the environment when manure is land applied as fertilizer. Additionally, the disposal of dead animals may result in the release of nanomaterials present in the animal's body.

4.1.1.2. Research Activities

ORD will identify industries, processes, and products that have relatively high potential to release engineered nanomaterials into the environment. Existing literature will be evaluated to better understand the industries of importance and identify where gaps in information preclude a full assessment of emission/release points of concern. ORD will perform a systematic assessment of the production, use, and ultimate fate of nanomaterials to understand the potential for emissions/releases into the environment. A modified tool using life cycle principles will be developed to better understand which industries pose the greatest potential to emit/release nanomaterials of concern and to inform decision-makers about the overall impact of engineered nanomaterials. This effort will also include a series of assessments for the highest priority industry categories. Results from ORD workshops will be used to guide industry and nanomaterial selection for assessment. Comparative assessments will be produced to help inform decision-makers at what stage in the lifecycle of nanomaterials interventions could be used to avoid future environmental pollution. The recent report from the Woodrow Wilson Institute entitled: *Green Nanotechnology: Its Easier than You Think*, among other documents, indicates the need for life cycle research. According to the Project on Emerging Technologies, "Nearly 400 company-identified nanotechnology-based consumer products are on the market... This figure does not include more than 600 raw material and intermediate components and industrial equipment items used by nanotechnology manufacturers who participated in a survey by EmTech Research." (*Green Nanotechnology:It's Easier than you Think*, Woodrow Wilson International Center for Scholars, p.6.) This effort will be closely coordinated with other organizations, particularly OPPTS which is also generating data on nanotechnology industries.

This research can be used to inform the Agency, industry, and academia about potential proactive and "greener" approaches for manufacturing nanomaterials that are designed to prevent nanomaterial release into the environment. It could also be used as input for future thorough LCAs.

High-potential industries/processes

ORD will draw upon the latest literature, hold workshops, and interact directly with industry representatives to identify market trends for nanotechnology industries that utilize the priority engineered nanomaterials indicated earlier in this document. This research will attempt to quantify the amounts of nanomaterial expected to be produced and used by existing industries, identify key processes used to manufacture these nanomaterials, and project future industries where significant releases may occur.

Identification/characterization of potentially released materials

Once we know where the engineered nanomaterials may be released, it will be important to understand something about the characteristics of these materials to inform future transport, transformation, exposure, and health studies. The research will focus on whether the nanomaterial emissions/

releases have the same characteristics (size, chemical composition) as the original material or have been modified before release to the environment. This area of research will be highly dependent upon the availability of technology to identify and characterize engineered nanomaterials. Unfortunately, the ability to make these measurements is also highly uncertain and will require extensive research. Efforts to identify, develop, test, and verify detection technologies will be critical to thc succcss of this research activity.

Entry point into the environment

Given that during the manufacture, use, and recycling or disposal of conventional products there are always emissions/releases of pollutants, it is reasonable to presume that some form of engineered nanomaterials will follow similar entry points into the environment. One of the primary goals of this research is to generate the data and tools needed to quantify and project these points of entry, so they can evaluate potential risks and possible approaches to manage those risks. One of the key issues to investigate is whether the nanomaterial compounds will be emitted/released in their original form or whether they will be physically or chemically bound with other compounds. This will directly impact transport and transformation and will influence potential exposures and health risks. Nanomaterials that are introduced to the environment in solution are more likely to remain in their original form and become bioavailable. Nanomaterials that are chemically cross-linked in a matrix are less likely to be released in their original form and size, although uncertainties remain. Because of their exceptional properties and characteristics, some engineered nanomaterials are being intentionally released to serve as catalytic agents for remediation or filtration purposes or as instruments for detection of pollution. This research will summarize the latest uses and provide available information on the characteristics of the materials released.

The goal of this research question is to perform the key initial step to inform additional research on transport, transformation, and subsequent exposure and health studies. In addition, by identifying potential release points, this research will provide key data required to inform how best to manage any potential risks.

4.1.1.3. Anticipated Outcomes

- Identification of industries, processes, and products that have relatively high potential to release engineered nanomaterials into the environment by working collaboratively with other organizations to inform decision-makers about the overall impact of engineered nanomaterials.
- Improved understanding of the industries of importance and identication where information gaps that preclude a full assessment of emission/release points of concern
- A systematic assessment of the production, use, and ultimate fate of nanomaterials that will improve our understanding of the potential for emissions/releases into the environment.
- Development of a modified tool using life cycle principles to: (a) better understand which industries pose the greatest potential to emit/release nanomaterials of concern and (b) inform decision-makers about the overall impact of engineered nanomaterials
- A series of assessments for the highest priority industry categories, the results of which will be used to guide industry and nanomaterial selection for assessment.
- Development of comparative assessments to help inform decision-makers at what stage in the lifecycle of engineered nanomaterials interventions could be used to avoid future environmental impacts.

4.1.2. Key Science Question 2: What technologies exist, can be modified, or must be developed to detect and quantify engineered nanomaterials in environmental media and biological samples?

4.1.2.1. Background/ Program Relevance

The detection of engineered nanomaterials in various environmental media presents a significant challenge. This is due in part to potential confounding by the presence of anthropogenic and natural nanomaterials. Challenges arise because many different engineered nanomaterials currently exist and their numbers are increasing exponentially; for certain types of nanomaterials, such as nanotubes, many thousands of different structures are possible. In addition, the fate, transformation, and mobility of these materials are only beginning to be understood. Consequently, scientific understanding of the reactions these materials undergo, how they age in various environmental media, how they interact with other compounds present in the environment, and whether and to what extent they form agglomerates or aggregates is limited. These issues compound the complexity of detecting and quantifying nanomaterials in environmental media.

The development of effective methods for measuring engineered nanomaterials in environmental media at concentrations relevant to potential exposure scenarios is critical to understanding the environmental impacts of these materials. Such methods would also enable the more rapid achievement of the safe development of nanotechnology-related products. ORD-sponsored research will ultimately seek to develop remote, in situ, and continuous monitoring devices that yield real-time information and that can detect engineered nanomaterials at very low concentrations.

Risk assessments of nanomaterials will require the ability to measure their environmental concentration in the workplace, home, biota (including human tissues), and ecosystems of interest. Analytical methods needed to characterize and analyze nanomaterials will require the modification of existing analytical tools and the development of completely new tools and approaches to meet these challenges. The same properties that make nanomaterials a significant challenge to analyze in any matrix (such as high binding capacities) may also provide unique opportunities for developing new analytical methods (e.g., tagging with fluorophores) for their analysis in complex biological and environmental systems. ORD will integrate fundamental research on detection method development from NSF, National Institute of Standards and Technology (NIST), DoD, and others with its own focused methods research effort to inform this research question.

4.1.2.2. Research Activities

Measurement science (based on analytical chemistry and physical properties) will have multiple roles in nanomaterials assessment and will require different types of analytical methods. There are several major areas of investigation with nanomaterials that require the application of a wide array of measurement and characterization techniques for characterization, detection, identification, or quantification:

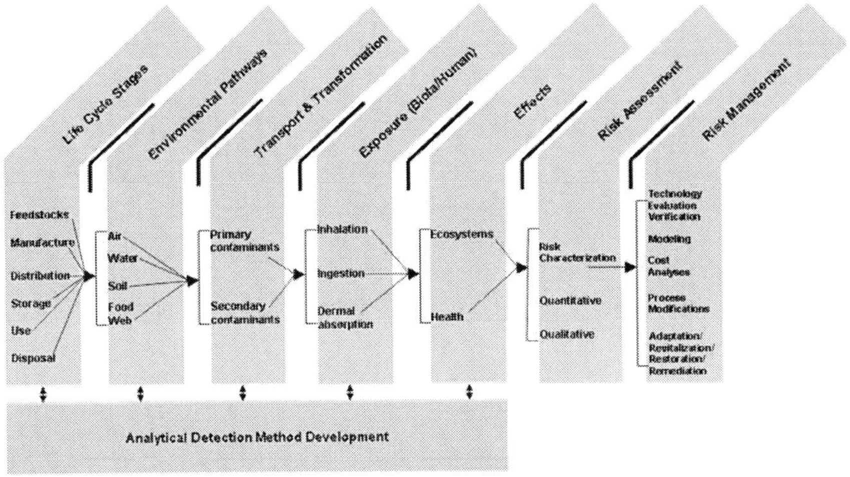

Figure 4-2. Relationship of Key Science Questions to Support Risk Assessment and Management Decisions – Analytical Detection

Bulk materials: ORD will undertake studies to characterize the physical and chemical properties of bulk nanomaterials to assess and quantify their unique features and characteristics (e.g., surface-to-volume ratio, 3-dimensional structure, size, size distribution, relative dimensions (aspect ratio), chirality, electrical/magnetic properties, and microstructure). Access to the equipment needed for these studies will require the formation of partnerships with other federal agencies, such as NIST, the National Cancer Institute (NCI) and the DOE. Each of these agencies has or is in the process of establishing nanomaterial research facilities, such as the Advanced Measurement Laboratory (AML) at NIST and the Nanotechnology Characterization Laboratory (NCL) at NCI. These research facilities provide access to a wide variety of measurement and characterizations tools.

Lab-based studies: ORD will take advantage of existing analytical methods for nanomaterials to support the initial focus on lab-based studies. An ever increasing number of papers in the literature have reported on the application of analytical methods for the measurement of nanomaterials for monitoring lab-based studies to model environmental processes under controlled conditions (e.g., soil leaching and subsurface transport) and concentrations. Examples include the analysis of fullerenes by liquid chromatography coupled to a photodiode array detector, the tracking of ^{14}C in radio-labeled carbon-based nanotubes, and the analysis of quantum dots by fluorescence spectroscopy.

Trace environmental residues: The published literature on the use of existing analytical tools for detecting or monitoring engineered nanomaterials in the environment (especially in matrices other than the vapor phase) is very limited. Perhaps the first publication that borders on being a "review" of this literature is that of Nowack and Bucheli (2007). The lack of methodologies for analyzing environmental samples likely results from two major factors: (1) only in the last couple of years has any need for environmental analysis been contemplated, and (2) the challenges facing the detection and quantification of engineered nanomaterials (especially those based solely on carbon) in environmental samples far exceed those associated with conventional pollutants, even those pollutants that comprise complex mixtures of many congeners (e.g., toxaphene).

To address the challenges associated with directly measuring nanomaterials in the environment, ORD will develop direct and indirect methods that capitalize on properties that are unique to these substances. For example, creating opportunities for indirect detection could capitalize on the extreme capacity of carbon-based nanomaterials to sorb certain chemicals, especially those sorbates that would be amenable to fast and sensitive detection. This could be done, for example, by equilibrating the sample unknown with an inorganic substance with strong sorptive potential. This substance would act as a dopant, which would be selected for its preferential sorption to carbon-based nanomaterials, its ready detectability, and the fact that it should rarely occur in the environment (to minimize background interference).

The complexities faced by analysis for nanomaterials in environmental matrices may prove intractable to conventional instrumented approaches of analysis. The eventual solution may well evolve from the development of new analytical approaches using arrays of standardized assays based on biological/biochemical endpoints. A battery of suitable assays could possibly be designed around a series of critical, evolutionarily conserved biological processes that prove keys to significant biological effects known to be important for nanomaterials. Two examples are: (1) the extent of physical penetration of a biological membrane (or membrane model) by the substances in a given sample (this would possibly be relevant to nanotubes), and (2) the generation of reactive oxygen species (as an indirect indicator of surface-catalyzed reactions). These endpoint assays would need to be developed to cover the entire spectrum of mechanisms of action for anthropogenic nanomaterials. Positive responses from these assays could then be used to direct the use of instrumented detection techniques to better target conventional analysis.

To make unambiguous and quantitative determinations of engineered nanomaterials in environmental samples, ORD will develop a combination of ensemble techniques (e.g., hyphenated methods coupling separation with spectroscopic detection, that measure collectively a number of particles) and single-particle techniques (e.g., methods, such as imaging, that measure individual particles). The separation method employed may be size exclusion chromatography, sedimentation field flow fractionation, or capillary electrophoresis. Determination could then be made by the coupling of ICPMS or a spectrofluorometer for fluorescent quantum dots. Ensemble methods can be developed for at least some classes of nanomaterials that provide screening assays to confirm the absence of detectable levels of nanomaterials or to provide an upper limit concentration estimate. However, for most materials, non-specific, indirect detection techniques would have to be combined with nanoscale imaging methods to confirm the presence of nanomaterials, and to provide a more reliable concentration estimate.

To develop analytical methods suitable for environmental monitoring, ORD will develop standardized reference materials in a variety of representative matrices. Methods for environmental analysis or routine monitoring must account for the extraordinarily wide array of potential parent materials and transformation products. In contrast to methods for the other roles described above, approaches to environmental measurement must include non- target analysis, where the type(s) of nanomaterials that need to be detected are not known in advance (the entire spectrum of parent materials must be amenable to analysis). The problems that traditionally plague environmental analysis, such as the wide array of matrix interferences that limit detectability, make environmental monitoring of nanomaterials even more challenging. Examples of this type of application do not yet exist, and are an additional research need.

4.1.2.3. Anticipated Outcomes

- Development of methods for characterizing nanomaterials, through partnerships with NIST, NCI and/or DOE
- Development of analytical methods for the detection of carbon-based nanomaterials in environmental matrices

- Development of analytical methods for the detection of non-carbon-based nanomaterials in environmental matrices
- In cooperation with other federal agencies, development of standardized reference materials for a variety of representative environmental matrices

4.1.3 Key Science Question 3: What are the major processes/properties that govern the environmental fate of engineered nanomaterials, and how are these related to physical and chemical properties of those materials?

4.1.3.1. Background/Program Relevance

Given the current scientific uncertainty surrounding fate, transport, detection and modeling of engineered nanomaterials, it is difficult to accurately assess the environmental disposition of nanomaterials or the potential exposure pathways to human and ecological receptors. Ultimately predictive models for estimating the environmental fate and transport of nanomaterials are needed.

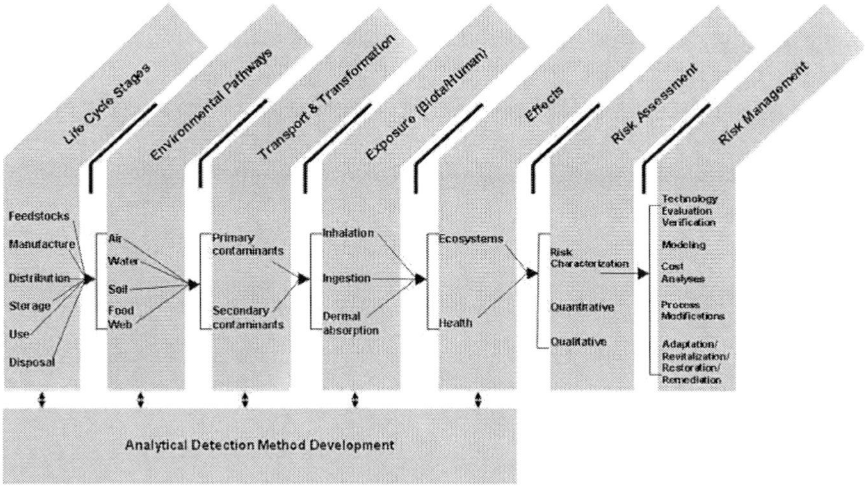

Figure 4-3. Relationship of Key Science Questions to Support Risk Assessment and Management Decisions – Pathways, Transport, and Transformation.

Nanotechnology research for fate, transport, detection and modeling of engineered nanomaterials is needed to identify the most critical parameters and uncertainties associated with these materials. This research will characterize the fate and transport of nanomaterials from sources to human and ecological receptors. The research will support risk assessments of engineered nanomaterials and ways to manage their potential releases. It will also provide a fundamental understanding of the physical and chemical properties of nanomaterials and their impact on fate and transport pathways. In concert with the research questions above, this research will also address detection issues of nanomaterials as it relates to fate and transport questions. Finally, existing predictive models for nanomaterial fate and transport will be modified, and if necessary, new models will be developed.

Because of the introduction and increased production of nanomaterials, it is necessary to better understand the fate, transport, detection and modeling of these materials. Quantitative as well as qualitative research is necessary to reduce the uncertainty surrounding the introduction and existence of nanomaterials in the environment and to identify the exposure pathways of concern to receptors. Research on these issues will assist the Agency in both risk assessment and risk management of engineered nanomaterials. ORD will conduct the following research to meet the critical needs of the agency as described below.

4.1.3.2. Research Activities

- Understand the processes that govern the fate and transport of engineered nanomaterials
- Understand the chemical and physical properties of engineered nanomaterials and how they influence fate and transport processes
- Develop predictive models for transport of engineered nanomaterials

Understand the processes that govern the fate and transport of engineered nanomaterials

ORD will work in collaboration with other agencies and academia to study the principles that govern the transformation, transport, and longevity of engineered nanomaterials in the environment. Since these materials could be present in sediments, soils, air and aqueous environments, understanding their transport in porous and compacted media is important to assess their migration through soils, the vadose zone, sediments, groundwater, surface water, and the atmosphere to potential receptors, as well as to develop effective management strategies. Studying the fate and transport of nanomaterials in all of these matrices is an important research need. Processes that control movement, sorption, dispersion, agglomeration, degradation, chemical and biological processes, and interactions between nanomaterials and natural or anthropogenic chemicals need to be investigated. ORD will conduct controlled laboratory studies to understand these fate and transport processes and the factors that control them.

Understand the physical and chemical properties of engineered nanomaterials and how they influence fate and transport processes

ORD will work in collaboration with other agencies and academia to study the chemical and physical properties of engineered nanomaterials and how these properties affect the fate and transport processes. Processes that control movement, sorption, dispersion, agglomeration, degradation, chemical and biological processes, are strongly affected by the chemical and physical properties of nanomaterials such as surface charge, pH, ionic strength, redox conditions, and ambient air conditions such as temperature and humidity. Obtaining information on the chemical and physical properties of specific nanomaterials and classes of materials is necessary to understand their effect on fate and transport processes. For example, in the case of carbon nanotubes, the mobility of these materials largely depends on the degree and type of functionalization (elements or other functional groups at the surface of the nanostructures), which affect solubility and surface charge. Research will focus on the understanding the impact solution chemistry and surface functionalization of multi-walled carbon nanotubes have on mobility in porous media.

Determining how transport through soils, vadose zone, and groundwater is affected by solution chemistry and colloid surface properties is critical for understanding the fate of nanomaterials. In addition, previous metals research has shown that chemical speciation of inorganic, engineered nanomaterials is an important factor to understand for the fate and transport and ultimate bioavailability of the materials. ORD will assess the chemical transformation and speciation of inorganics such as silver. (Silver is impregnated in fabrics and washing machines as an anti-fungal/anti-microbial agent, but little is known about how the properties of the nanosilver particles impact their fate and transport in the environment.)

Develop predictive models for transport of engineered nanomaterials

ORD will work in collaboration with other agencies and academia to study the applicability of existing environmental fate and transport (EF&T) models and to develop new predictive EF&T models that are tailored specifically to nanomaterials. Early analysis of the Estimation Programs Interface Suite (EPI Suite) models, the primary set of predictive tools the Agency uses for calculating the fate and transport of soluble organic chemicals and inorganics, indicates that they will have little or no applicability to predicting the EF&T of nanomaterials. Models do exist for predicting the transport of larger particle sized colloidal materials and they are being investigated for application to nanomaterials. As such, traditional DLVO (Derjaguin, Landau, Verwey and Overbeek) theory will likely lend insight into environmental fate and mobility trends. The successful development of EF&T models for nanomaterials will depend on our understanding of the processes controlling the EF&T of engineered nanomaterials and our ability to determine the chemical and physical properties needed to predict such processes.

4.1.3.3. Anticipated Outcomes

Results from this research will provide an improved understanding of the EF&T of engineered nanomaterials in the environment. This will allow the Agency to develop a set of predictive tools.

Researchers hope to:

- Develop a scientific understanding of the processes that govern the fate and transport of engineered nanomaterials
- Measure the chemical and physical properties of engineered nanomaterials and determine how these properties influence and impact fate and transport
- Identify the exposure pathways associated with production, end-use, and recycling or disposal of engineered nanomaterials in different environmental matrices
- Improve the scientific understanding of detection methodologies for quantifying engineered nanomaterials
- Develop multiple predictive models for understanding and measuring the transport of engineered nanomaterials

4.1.4. Key Science Question 4: What are the exposures that will result from releases of engineered nanomaterials?

4.1.4.1. Background/ Program Relevance

Research is needed to provide insight into the type, extent, and timing of exposures to nanomaterials in all relevant environmental media and through all relevant exposure pathways. Cumulative exposures, both with other engineered nanomaterials as well as with bulk-scale pollutants, also need to be explored. The information provided through this exposure research can be linked with other exposure and biological impact data to improve the scientific basis of risk assessment for engineered nanomaterials.

General population exposure may occur from environmental releases from the production and use of nanomaterials and from direct use of products (e.g., cosmetics and medicines) containing nanomaterials. The rapid growth of products that contain nanomaterials could result in their presence in soil and aquatic ecosystems. This presence will result from effluents of manufacturing plants, and the recycling or disposal of nano-based consumer products into landfills and surface/ground water.

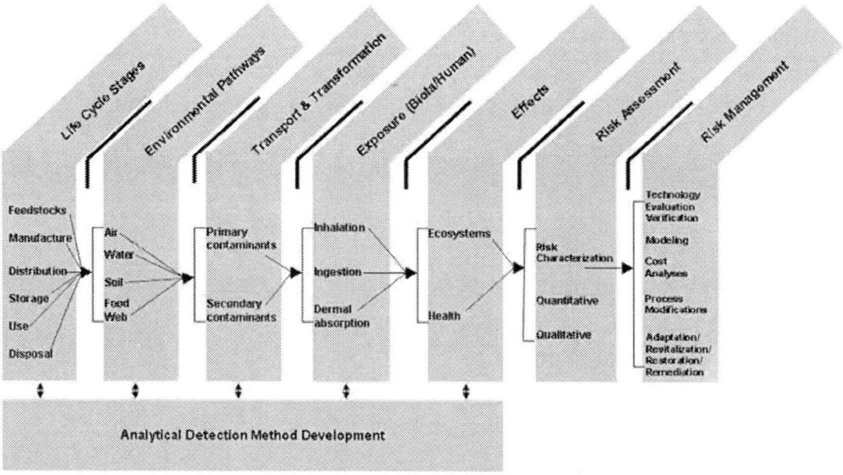

Figure 4-4. Relationship of Key Science Questions to Support Risk Assessment and Management Decisions – Exposure.

An exposure assessment attempts to answer the following questions for a particular substance or chemical:

- Who or what is exposed (e.g., people, aquatic ecosystems)?
- What are the pathways for exposure?
- How much exposure occurs?
- How often and for how long does exposure occur; that is, what is its frequency and duration?

4.1.4.2. Research Activities

The Agency uses a number of models to conduct chemical exposure assessments. Descriptions and links to these models can be found at the websites for the Council for Regulatory Environmental Modeling (CREM: http://cfpub.epa.gov/crem/) and the Center for Exposure Assessment Modeling (CEAM: http://www.epa.gov/ceampubl/). Table 4-1 provides a listing of several of the models/tools used by the program offices for exposure assessment and each model's general application and applicability to nanomaterials in its current form. With the exception of the EPI-SuiteTM calculators, all of the exposure assessment models need the user to provide input data on the physical and chemical properties for the chemical of interest. The EPI Suite™ calculators are based on a single input, a Simplified Molecular Identification and Line Entry System (SMILES) string that is a typographical method for representing unique chemical structures. The other models in Table 4-1 were developed primarily for exposure assessments of synthetic organic chemicals, and thus require input such as water solubility, octanol-water partition coefficients and Henry's Law constants to predict fate and transport.

Table 4-1. Several of the primary models/tools used by the Program Offices for exposure assessment and each model's general application and applicability to nanomaterials in their current form

Exposure models will require modification to allow the input of molecular parameters and physical and chemical data specific to nanomaterials (e.g., particle size, surface charge, distribution or sticky coefficients, and agglomeration tendencies). OPPT has recently requested the assistance of ORD to review the E-FAST model, which supports the New Chemicals and Existing Chemicals Programs, for its applicability to nanomaterials. Specifically, ORD will:

Acronym	Model Name	Primary Program Office	Application	Applicability to NMs
E-FAST	Assessment Screening Tool Version 2.0	OPPT	Estimates concentrations of chemicals in multi-media from multiple release activities	Modification Required
EPI-SuiteTM	Estimation Programs Interface Suite	OPPT	Estimates physical & chemical properties for organic chemicals	Not Applicable
EXAMS	Exposure Analysis Modeling System	OPP	Estimates fate, transport, and exposure concentrations of chemicals in aquatic ecosystems	Modification Required
Trim Expo	Total Risk Integrated Methodology Exposure-Event Module	OAQPS	Estimates human exposure to criteria and hazardous air pollutants	Modification Required

- Focus on the physical, chemical, and other properties currently required as user provided/default inputs
- Determine whether these inputs are appropriate for nanomaterials when assessing exposures related to industrial releases to surface water, air, and/or landfills
- Identify other properties as potential inputs that might be more appropriate for assessing general population and environmental exposure to nanomaterials

The challenges in identifying and measuring the concentration of engineered nanomaterials in environmental and biological systems will present significant obstacles to providing the data necessary to conduct exposure assessments of these materials for both ecological and human receptors. Such assessments will require the development of alternative methods for determining the source and the environmental concentrations of nanomaterials in aquatic and terrestrial ecosystems. The interest in nanomaterials is driven by their unique properties and activities at different scales; these same properties provide the opportunity for developing indicators of exposure by measuring changes in structures and functions of biological organisms in contact with nanomaterials. By identifying indicators of exposure resulting from exposure to nanomaterials, it will be possible to reconstruct the exposure pathway and ultimately the source and the environmental concentration of the nanomaterial of interest. This ability to move from an internal biological response to external environmental concentration represents a growing area of exposure science referred to as "Exposure Reconstruction."

ORD' s research in this area focuses on the linkage of responses across endpoints at multiple biological levels of organization, from molecular alterations to populations. These linkages can serve as a basis for identifying and validating mechanistic indicators of exposure and effects, informing ecological risk assessments of nanomaterials. Currently, a systems-based approach is being used to assess exposures and define toxicity pathways for model chemicals with well-defined modes/mechanisms of action (MOA) within the hypothalamic-pituitary-gonadal (HPG) axis. These pathways serve as a basis for understanding responses of small fish across biological levels of organization, ranging from molecular responses to adverse effects in individuals to, ultimately, changes in population status. The studies employ a combination of state-of-the-art molecular biology, bioinformatic, and modeling approaches, in conjunction with whole animal testing. As such the project will enable a unique opportunity to interface empirical toxicology with computational biology in the exposure assessment of nanomaterials.

The molecular biological tools for this research will focus on the application of the 'omic' tools (i.e., genomics, proteomics and metabolomics) to identify indicators of exposure. These tools provide the ability to identify indicators of exposure by measuring gene regulation, protein formation, and changes in an organism's metabolome in response to exposure to a chemical or mixture of chemicals. By elucidating the kinetics of the marker's response, it is also possible to provide an understanding of the temporal and spatial aspects of exposure.

Currently, no information is available in the literature concerning the identification of indicators of exposure for nanomaterials. On-going research with pesticides exhibiting estrogenic activity, however, is demonstrating the feasibility of this approach. ORD has developed molecular indicators of exposure (based on genomic responses) of aquatic organisms (water flea, *Daphnia magna* and fathead minnow, *Pimephalespromelas*) to estrogenic compounds and is using advanced genomic methods to develop androgenic indicators. The Nuclear Magnetic Resonance (NMR) based metabolomics research program being conducted at ORD's NMR research facility is demonstrating the use of high-resolution NMR to identify changes in the profiles of endogenous metabolites (i.e., the metabolome) in the serum and urine of fathead minnows exposed to estrogenic compounds. The literature also provides examples of the use of genomics to identify indicators of exposure in humans. Microarray analysis of blood samples taken from benzene-exposed workers has identified peripheral blood mononuclear gene expression as an indicator of exposure for benzene (Forest et. al, 2005).

Collaboration to further identify the exposure pathways of engineered nanomaterials

ORD will work in collaboration with other agencies and academia to study and identify the most common exposure pathways for engineered nanomaterials. ORD will seek to establish international collaborations through the development of collaborative or coordinated calls for proposals. These research proposals will also engage ORD scientists in the study of exposure routes and pathways, relevant exposure doses, and critical exposure concentrations. Research will also identify potential subpopulations of organisms that are more susceptible to engineered nanomaterial exposure than others.

4.1.4.3. Anticipated Outcomes

- Identification of the dominant exposure pathways to ecological receptors of interest
- An assessment of the applicability of the Agency's current exposure models to nanomaterials
- Identification of the physical and chemical properties required to inform exposure
- Identification of indicators of exposure through the application of genomics, proteomics and metabolomics

4.2. Research Theme: Human Health and Ecological Effects Research to Inform Risk Assessments and Test Methods

4.2.1. Key Science Question 5: What are the effects of engineered nanomaterials and their applications on human and ecological receptors, and how can these effects best be quantified and predicted?

4.2.2. Background/Program Relevance

As described in EPA's Nanotechnology White Paper, nanomaterials could have health and ecological implications arising from new routes of exposure and/or toxicities associated with either direct exposure to these novel materials, by-products associated with their applications, or their interactions with various environmental media.

By understanding nanomaterials biokinetics, characterizing their health and ecological effects, and identifying the physical and chemical properties that regulate their toxicity, ORD will address the critical lack of information required for nanomaterials risk assessment. The results from ORD's nanomaterials health and ecological effects research will also inform risk management strategies and decisions.

ORD's health and ecological effects research will provide EPA offices with information on the health and ecological effects of specific nanomatierals and their applications, as well as guidance on best practices and approaches/test methods for assessing/predicting health and ecological effects. ORD will also be addressing key immediate priority effects research needs identified in US EPA Nanotechnology White Paper, such as, adequacy of test methods, characterization of the health effects of nanomaterials (nanotoxicology), hazard identification and dosimetry and fate.

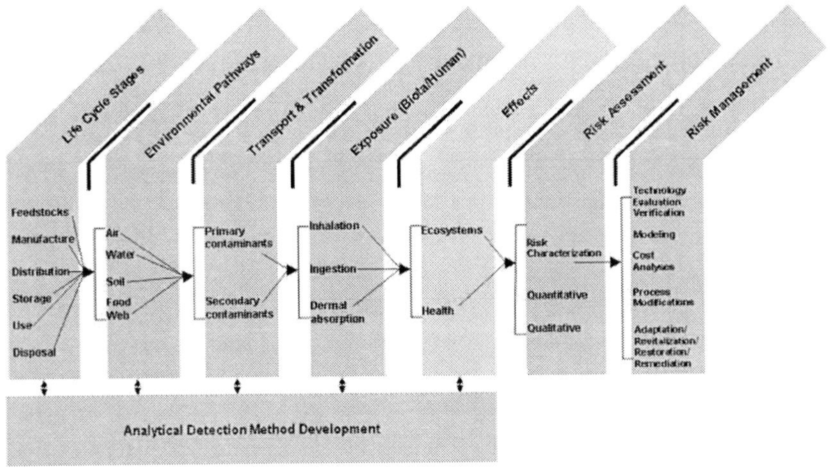

Figure 4-5. Relationship of Key Science Questions to Support Risk Assessment and Management Decisions – Effects

4.2.3. Research Activities

To address this key science need, ORD will conduct research to:

- Evaluate current test methods to assess their adequacy to determine the toxicity of nanomaterials, and develop new toxicity test methods, as required
- Determine the health and ecological effects of nanomaterials, including acute and chronic effects and local and systemic effects
- Determine the health and ecological effects associated with nanomaterials applications and/or interaction with environmental media, ecosystems, or other stressors
- Determine if toxicity, mode(s) of action, and mechanism(s) of injury are unique to the novel physical and chemical properties of nanomaterials
- Identify factors and properties regulating deposition, uptake, fate, and toxicity of nanomaterials (including hazard identification; dose-response correlations; ADME; and susceptibility/sensitivity host factors)
- Identify ecological systems that contain especially susceptible organisms, life stages, or populations
- Develop alternative approaches/technologies/models to screen, rank, and predict the *in vivo* toxicity of nanomaterials and their applications

Health effects: ORD's nanomaterial health and ecological implications research builds upon its ongoing risk assessment research within the Air, Water, and Safe Products/Safe Pesticides programs. These research activities provide the facilities and expertise that are directly applicable to addressing the health and ecological implications of nanomaterials resulting from various potential routes of exposure. ORD's research is conducted within a risk assessment paradigm to address key research issues listed above.

A Multi-tiered strategy for assessing nanomaterial health effects: To address the nanomaterial health and ecological research needs identified in EPA Nanotechnology White Paper (1), Environmental, Health and Safety Research Needs for Engineered Nanoscale Materials (2), and be consistent with recommendations of the National Academy of Sciences, National Research Council report on Toxicity Testing in the 21st Century: A Vision and a Strategy (3), ORD's research will employ a multi-tiered strategy, Figure 4-6.

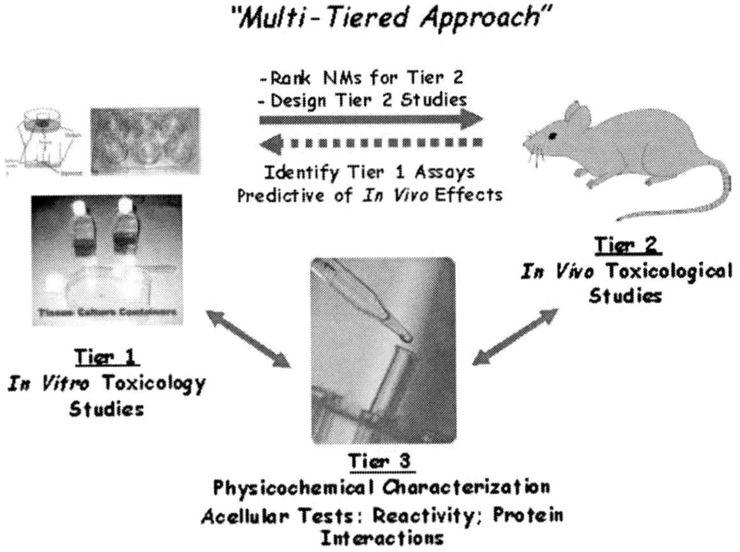

Figure 4-6. A multi-tier strategy for comparative and quantitative nanomaterials health risk assessment.

ORD's nanomaterials health effects multi-tiered strategy is driven by a number of additional critical factors such as: the diversity of engineered/manufactured nanomaterials; the cost and availability of nanomaterials; the need to identify alternative approaches, assays, and methods that predict *in vivo* health effects resulting from direct exposure to nanomaterials or following their interactions with environmental media resulting from inadvertent releases or applications.

Tier 1 - In Vitro Toxicology of nanomaterials: Initially, studies will examine the *in vitro* toxicity of various nanomaterials of interest to the Agency using a variety of cell types reflecting different routes of exposure (inhalation, oral, dermal) to assess the health effects that may arise due to the tendency of nanomaterials to translocate to other regions of the body. This "virtual body" approach employed in Tier 1 studies will assess the *in vitro* cancer, pulmonary, immunological, neurological, reproductive, cardiovascular, and developmental toxicities of nanomateials. Tier 1 *in vitro* testing provides a means to: rapidly screen and rank the relative toxicities of various nanomaterials; determine mechanism(s) of injury and mode of action; rapidly perform comparative toxicity studies between nano vs. bulk size materials; conduct rapid screening to assess alterations in nanomaterials toxicity following their interactions with environmental media; perform ADME at the cellular and intracellular levels; and characterize nanomaterials–cellular interactions.

ORD's ToxCast program (http://epa.gov/comptox/toxcast/news.html) will assist ORD's Tier 1 *in vitro* toxicological assessment of nanomaterials. ToxCast offers an approach to deal with the extreme diversity of nanomaterials by applying high-throughput platforms and computational approaches (physicochemical properties, biocomputational models, biochemical assays, cellular assays, genomic studies, and model organisms) to screen a large number of materials. The ToxCast program has the potential to rank the toxicity of nanomaterials as well as develop models to identify physical and chemical properties that determine the toxicity of nanomaterials.

Tier 2 – In Vivo Toxicology of Nanomaterials: Subsequent Tier 2 studies will examine the animal or *in vivo* toxicity and biokinetics/ADME of nanomaterials. Tier 2 studies will be guided by information generated in Tier 1 related to the prioritizing or ranking of nanomaterials and designing appropriate nanomaterial exposure concentrations as well as what health endpoints to monitor, Figure 4-6, solid red line. Tier 2 studies will examine cancer, pulmonary, dermal, and gastrointestinal toxicities associated with initial deposition of nanomaterials by various routes of exposure as well as immunological, neurological, reproductive, cardiovascular, and developmental toxicities to assess their systemic toxicities. Information generated from Tier 2 studies will provide a database from which to compare Tier 1 studies in order to identify those *in vitro* assays that correlate with *in vivo* nanomaterial toxicity or health effects, Figure 4-6, dashed red line.

Tier 3 – Nanomaterial Characterization and Surface Properties: Concurrent with Tier 1 and 2 activities, Tier 3 research will relate the physical and chemical properties of nanomaterials to their *in vitro* and *in vivo* toxicity (hazard identification), Figure 4-6, red solid double-headed arrows. Tier 3 research will employ non-cellular or acellular methods to assess nanomaterial surface reactivity as well as understand their interactions with biological molecules/fluids in order to identify what surface properties and interactions determine their *in vitro* and *in vivo* biokinetics/ADME. These studies will also investigate nanomaterials effects in a variety of cell types and organ systems.

The multi-tier strategy will not only provide an approach to perform comparative and quantitative nanomaterials health effects risk assessment for a number of different types of nanomaterials of interest to Agency offices, but also offers an approach to assess alterations in NM toxicity following their interactions with environmental media. Critical components of Tiers 1, 2 and 3 are the use of high throughput screening assays and the application of "omic-based" analyses and associated bioinformatics to characterize the health effects and molecular response profiles. This research may lead to the identification of biomarkers of exposure and/or effects as well as the identification/validation of *in vitro* toxicity and acellular test methods that predict the *in vivo* toxicity of nanomaterial s.

Ecological effects

Tier 1 - Evaluate the suitability of existing test methods for assessing the hazards of engineered nanomaterials: Nanomaterials, or products containing nanomaterials, are already being submitted for approval under Agency programs such as TSCA and FIFRA. These and other Agency programs have existing protocols for evaluating hazards to ecological receptors in both aquatic and terrestrial systems, but the appropriateness of these methods for nanomaterials has yet to be evaluated. Key concerns include how to expose organisms to nanomaterials in ways that have relevance to exposures that may occur in the environment, and whether these standardized assays address the organisms, life stages, and bioavailability considerations that are most important for understanding the potential ecological risks of nanomaterials. In addition to direct toxicity testing, emphasis will be placed on measurements of exposure, uptake, and dose.

Tier 2 - Understand the mechanisms underlying the ecological effects of nanomaterials and identify potential gaps in hazard assessment procedures: Building on results of exposures using standard (or appropriately modified) test methods, further research will explore the specific mechanisms of nanomaterials toxicity and ecological effects. Understanding the mechanisms of effects is key to determining novel risks that may be created by nanomaterials, defining the appropriate organisms and endpoints for nanomaterials risk assessments, and providing the basis for future predictive models. Parameters that govern adsorption, distribution, metabolism, and excretion (ADME) will be evaluated, as will means of expressing toxicological dose. Other studies will evaluate the interaction of nanomaterials with physical, chemical, and biological components of ecological systems to determine if there are effects of nanomaterial s not captured by single organism toxicity testing, such as altering the relationships among ecosystem components and thereby affecting overall ecosystem function. Throughout Tier 2, emphasis will be given to determining whether nanomaterials exert effects through mechanisms that would not be well addressed by existing ecological hazard and risk screening tools.

Tier 3 - Development of methods and models to predict the hazard or ecological risk of nanomaterials: Due to the diversity of nanomaterials expected to enter the marketplace in the coming years, the Agency will need predictive tools that can be used to prioritize newly developed nanomaterials for testing and further evaluation. For example, quantitative structure/activity relationships (QSARs) may be developed to predict the toxicity of untested materials based on their chemical structure and an understanding of the mechanisms underlying dose and toxicity. Likewise, ecological effects models may be important predictive tools if research in Tier 2 indicates that ecological processes above the organismal level are being uniquely affected by nanomaterials. This work will build directly from Tiers 1 and 2 and associated research conducted by the Computational Toxicology Program.

Leveraging research with ORB laboratories, centers and other federal programs: ORD's nanomaterial health and ecological risk assessment research will leverage work with other ORD Federal programs (NIOSH, NTP, DOE) where similar nanomaterials are being monitored, studied, and characterized. For example, ORD laboratories are jointly addressing nano-cerium dioxide assessing potential environmental exposures, and associated health effects. Research to examine the health and ecological effects of nanomaterials following their release into or interactions with environmental media will require the combined expertise of ORD's health and exposure scientists. Finally, the physical and chemical characterization of nanomaterials and their detection in biological systems will require a multidisciplinary approach with close interactions across ORD as well as the DOE National Laboratories.

4.2.4. Anticipated Outcomes

ORD's effects research will provide key information regarding the health and ecological implications from exposures to nanomaterials, and their applications, in order to identify and manage potential adverse impacts and inform program offices and regions regulatory and other policy decisions. Specifically, ORD's nanomaterials effects research will provide Agency offices with information on the health and ecological effects of specific nanomaterials and their applications, as well as guidance on best practices and approaches/test methods for assessing/predicting health and ecological effects. ORD's nanotechnology health and ecological effects research activities will provide publications in peer-reviewed scientific journals on the:

- Characterization of nanomaterials health and ecological effects; identification of physical and chemical properties and host/sensitivity factors that regulate nanomaterials dosimetry, fate, and toxicity
- Identification of testing methods/approaches to predict *in vivo* toxicity of nanomaterials; characterization of molecular expression profiles that may provide biomarkers of nanomaterial exposure and/or toxicity
- Provision of necessary counsel and guidance that will assist in the review of premanufacture notice applications and assess the adequacy of harmonized nanomaterial test guidelines to assist OPPTS and internationally, the OECD
- Addressing the gap in our knowledge regarding the toxicity of nanomaterials which has impeded the ability to conduct accurate life cycle analysis

4.3. Research Theme: Developing Risk Assessment Methods

4.3.1. Key Science Question 6: Do Agency risk assessment approaches need to be amended to incorporate special characteristics of engineered nanomaterials?

4.3.2. Background/Program Relevance

Many data gaps exist in the areas of chemical and physical identification and characterization, environmental fate, environmental detection and analysis, potential releases and human exposures, human health effects, and ecological effects. Filling these data gaps will aid in future risk assessments of nanomaterials when proven risk assessment methods are available.

Although nanomaterials have special properties that may influence their environmental behavior and effects on human health and ecosystems, the traditional paradigm for risk assessment and risk management (NRC, 1983) is presumed to apply to these materials. Hazard identification determines qualitatively whether the nanomaterial will cause an adverse health effect.

Dose-response assessments establish the quantitative relationship between dose and incidence of health effects. Exposure assessment is performed and the incidence of the adverse effect (risk) in a particular population is determined by combining exposure and dose-response. The effects of nanomaterials on the environment must also be assessed in order to protect and restore ecosystem functions, goods, and services. Ecological risk assessment entails the evaluation of goals and selection of assessment endpoints in a problem formulation step, followed by analysis of exposure to stressors and determining the relationship between stressor levels and ecological effects. The next step is estimating risk through the combination of exposure and stressor-response profiles, description of risk by discussing lines of evidence, and determination of ecological adversity (U.S. EPA, 1998). Interfacing among risk assessors, risk managers, and interested parties during the initial planning of a risk assessment and communication of risk at the end of the risk assessment are critical to ensuring that the results of the assessment can be used to support a management decision. The importance of constant communication and stakeholder involvement in both human health and ecological risk assessment and risk management has also been noted by the Presidential/Congressional Commission on Risk Assessment and Risk Management (see Figure 2-1).

While the basic paradigms of health and ecological risk assessment are still relevant, they are expanded in the comprehensive environmental assessment (CEA) approach to encompass the product life cycle of nanomaterials. By taking a broad view of the potential for releases of both primary and secondary materials to multiple environmental media, the evaluation of the environmental and health risks of nanomaterials is seen as an issue that cuts across EPA programmatic domains and is not simply categorized as solely an air, water, toxics, or solid waste issue. The CEA approach (Davis and Thomas, 2006; Davis, 2007) starts with a qualitative life cycle framework, as shown in Figure 4-7. It takes into consideration multiple environmental pathways, transport and transformation processes, cumulative and aggregate exposure by various routes, and ecological as well as human health effects. Depending on the availability of data, both quantitative and qualitative characterizations of risks may result. However, given the limited information currently available on nanomaterials, the CEA approach is being used to identify where key data gaps exist with respect to selected case studies of specific applications of nanomaterials.

Case studies are recommended in the EPA Nanotechnology White Paper as a means to further inform research supporting the risk assessment process.

The term "case study" is used to refer to specific examples of nanomaterials and the types of issues that would be need to be considered to assess their respective environmental and health risks. By focusing on specific examples of nanomaterials in realistic applications, it is possible to identify and prioritize research needs to assess the "real world" impacts of these materials. Given the striking differences in toxicological and physicochemical properties of nanomaterials, generalizations across nanomaterials need to be considered cautiously.

4.3.3. Research Activities

The role of ORD's nanomaterial risk assessment research is (1) to help guide overall research efforts toward generating the information needed to conduct future comprehensive environmental assessments of nanomaterials and (2) to carry out such assessments in coordination with all of ORD and the program offices. The research question ORD will address is "Do Agency risk assessment approaches need amended to incorporate special characteristics of engineered nanomaterials?" To answer this question, ORD will identify and prioritize information gaps by conducting a series of case studies and workshops to further refine research needs for specific nanomaterials, as recommended in the EPA Nanotechnology White Paper.

In order to develop case studies of particular nanomaterials and their specific applications, appropriate nanomaterials must be selected. The collective judgment of an internal workgroup representing all relevant program offices was used for this purpose. The workgroup was given a summary of available information on the chemistry, human health, toxicology, exposure, and release of various nanomaterials. Workgroup members were then asked to select two nanomaterials based upon five criteria: potential for biota/human exposure; apparent potential for both health and ecological effects; a reasonable amount of information with which to develop a case study; relevance of the nanomaterial to programmatic or regulatory needs; and "nanoness," i.e., satisfying the NNI definition of having at least one dimension less than 100 nm. Using these criteria, titanium dioxide and single walled carbon nanotubes were selected. Two applications of nanotitanium dioxide are under development, a water treatment agent and a sunscreen. The applications for the single walled carbon nanotubes have not yet been determined. These selected classes of nanomaterials also serve as a common focus and point of coordination for near-term studies by the various ORD laboratories.

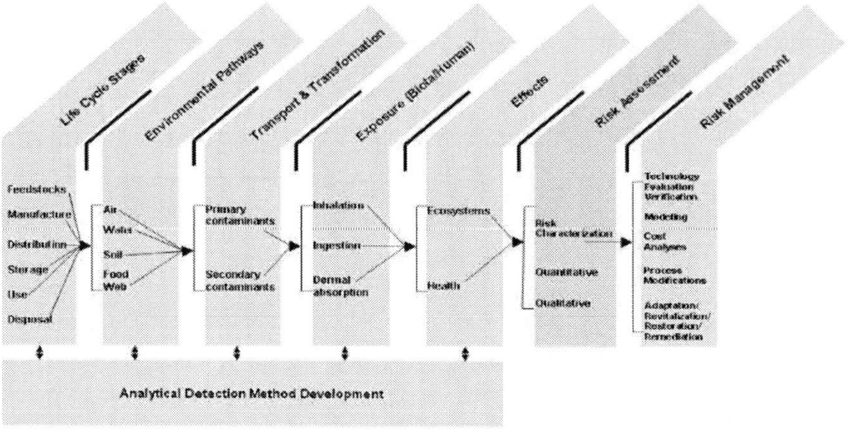

Figure 4-7. Relationship of Key Science Questions to Support Risk Assessment and Management Decisions – Risk Assessment

The intent of the case studies is to consider currently available information for nanomaterials for the purpose of identifying gaps where additional information is needed. The draft case studies will be internally reviewed, followed by distribution of each draft to selected reviewers/contributors as part of a peer consultation process. After further development and refinement through peer consultation, the case studies will be the subject of a workshop (likely the first of a series of such meetings) involving invited technical experts and stakeholders. The workshop will be conducted in a formal, structured manner using experienced facilitators trained in expert judgment techniques (e.g., multi-criteria decision analysis, expert elicitation). A detailed summary of the discussions and views expressed during the workshop will be used in refining the current research strategy document. This summary will highlight areas of work that will be needed to support comprehensive environmental assessments of nanomaterials. This refined statement of research directions will provide longer term guidance for both ORD and the broader scientific community.

4.3.4. Anticipated Outcomes

- Development of 3–4 draft case studies for specific applications of nano-titanium dioxide and single-wall carbon nanotubes. Each draft case study will undergo internal workgroup review.

- Administration of external peer consultation review, elaboration, and refinement of the draft case studies
- Scheduling of a workshop for invited experts and stakeholders and public observers, using formal expert judgment methods to identify and prioritize research needed to support comprehensive environmental assessments of nanomaterials
- Using input from the workshop discussions, a document that lays out long range research directions for obtaining information needed for nanomaterial CEAs

4.4. Research Theme: Preventing and Managing Risks

4.4.1 Key Science Question 7: What technologies or practices can be applied to minimize risks of engineered nanomaterials throughout their life cycle, and how can nanotechnologys' beneficial uses be maximized to protect the environment?

4.4.2. Background/Program Relevance

While it is critical to understand the potential environmental implications of nanotechnology, it is also important to investigate how various nanomaterials can be used to prevent, control or remediate environmental contaminants that have up to now been difficult to manage with conventional technology (Figure 4-8). Nanotechnology will be used to both create new technologies and improve the performance of conventional technologies. There are several avenues to obtain environmental benefits from nanotechnology.

- Use nanoscale materials in a synthesis process as a substitute for more toxic components or as a process mediator that reduces the mass of potentially toxic materials employed in the chemical process (e.g., catalysts)
- Incorporate nanoscale materials into a part of the production process used to treat noxious chemicals prior to final discharge
- Employ nanoscale materials to treat emissions/releases from power production and industrial processes waste streams
- Treat contaminated environmental media (i.e., air, water, sediments, or soil)

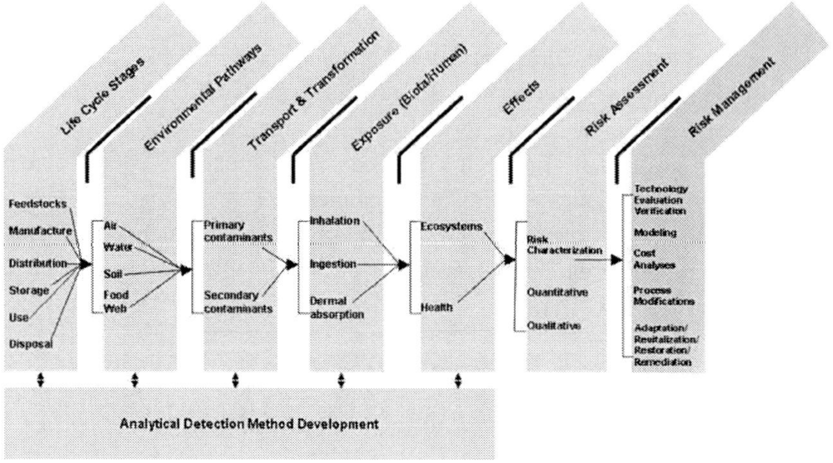

Figure 4-8. Relationship of Key Questions to Support Risk Assessment and Management Decisions – Risk management

ORD's initial emphasis will be to address key pollutants of concern to EPA program and regional offices that have historically been difficult to manage, including sources that emit low concentrations of air pollutants and remediation of hazardous materials in complex heterogeneous environments. The Woodrow Wilson Center 2007 report, "Green Nanotechnology: It's Easier than You Think," describes a variety of potential environmental benefits associated with use of the nanotechnology for environmental improvement.

In addition to supporting the recommendations of outside experts, this research will be valuable to EPA program and regional offices and outside stakeholders such as industry and states who are constantly looking for innovative solutions to address intractable pollution problems. Many of these needs (see Appendix A) have already been identified.

As the ORD research program progresses and identifies potential problems with specific engineered nanomaterials and products, risk management research will be directed to respond to study the impacts of these materials and products. This response could include process change recommendations that reduce/prevent the amount of engineered nanomaterials released/emitted or using the unique properties of nanomaterials to reduce potential risks.

A substantial increase in nanomaterial manufacturing is predicted in coming decades. When these particles or their nano-sized manufacturing or degradation byproducts find their way into water, land, and air, it will be necessary to effectively and efficiently remove or detoxify these substances. As a result, another key component of this science question will be to quantify how well technologies now in place reduce emissions/releases of potentially hazardous engineered nanomaterials. While these systems were not originally designed to capture such small materials or associated by-products, some technologies may be able to reduce them particularly if they have become bound to larger particles which the systems were designed to control.

ORD will conduct various workshops with industry, academia, and other parts of EPA to discuss potential environmental liabilities associated with manufacturing, using, recycling, and disposing of nanomaterials. The parties will exchange information and ideas about where releases are more likely to pose the greatest risks and what alternatives (e.g., preferred manufacturing approaches via green chemistry) are available that could minimize environmental liabilities. These workshops will help all participants consider how nanotechnology products can be designed in the most environmentally sustainable manner possible.

4.4.3. Research Activities

Research devoted to the capture of engineered nanomaterials or degradation by-products using conventional technology will address the ability of these technologies to manage releases of engineered nanomaterials to all media during their production. For nanomaterials that cannot be efficiently treated or controlled, this may indicate that production and use should be strictly controlled. Example abatement technologies to be evaluated include: primary, secondary and tertiary drinking water treatment plant technologies; best management practices (BMPs) for contaminated storm water and combined sewer overflow; wastewater treatment technologies; membrane technology; adsorption; and conventional particulate control technologies. The data collected will indicate whether existing abatement procedures or technologies are adequate or require substantial revisions to control nanomaterials.

This research will inform regulatory officials and industry about whether there are potential risks posed by the releases of engineered nanomaterials into the environment and what potential controls might be available to limit potential risks. This research has the potential to influence decisions regarding manufacturing, importing, storage, handling, and use of selected nanomaterials. The results of the research will be provided in the form of reports and computer-based systems that can be used to address the unique issues associated with various industrial operations.

Materials modification to support green manufacturing of nanomaterials

Research on greener synthesis approaches will identify opportunities to reduce the environmental implications of nanomaterial production. Since basic nanotechnology production processes are still under development, EPA is well placed to work with others to design production processes that minimize or eliminate any emissions/releases. The goal of this research will be to answer the question: how can energy consumption be minimized and waste/pollution prevented in the manufacturing of nanomaterials and products? The general approach will be to develop a strategy that allows the greener preparation of these materials. Three of the main green chemistry areas that will be investigated include: 1) the choice of solvent, 2) the reducing agent employed, and 3) the capping agent (or dispersing agent). For example, ORD is using a flame and furnace reactor combination to produce single-walled and multi-walled carbon nanotubes. Researchers are using a common feed-stock (e.g., propane), as opposed to mixtures of carbon monoxide and hydrogen, and a metallic catalyst to initiate nanotube formation. The challenges are to achieve high-quality and high-yield carbon nanotubes, and to use them for adsorption and catalyst support to enhance control of selectivity, activity, and stability.

Waste/byproduct minimization

The use of nanotechnology in industrial processes has many potential advantages. One potentially significant environmental benefit is reducing the amount of material sent to the waste stream. Under this research area, ORD will work with its partners in industry and academia to investigate advanced approaches that have the potential to reduce waste products in those industrial sectors with high volumes of waste. Waste minimization benefits to be realized through nanotechnology applications will result either through the substitution of less-toxic chemical components in the manufacturing process or through the reduction in the required mass of toxic chemical components via enhanced reaction rates or efficiencies. An example of the first scenario includes the use of nanomaterials to improve material characteristics of bio-based, nanocomposite products. These products are being developed as substitutes for more traditional petroleum-derived materials, resulting in a reduction of the mass of toxic components that could potentially be released into the environment. There are also numerous examples of the development of nanomaterials for use as catalysts in chemical manufacturing processes. The use of nanoscale catalysts results in an overall enhancement of process efficiency, thus reducing the required mass of toxic chemical components used in the manufacturing process.

Application of nanomaterials to reduce environmental risks

Under this research area, ORD will investigate the potential for various nanomaterials to minimize the release of toxic chemical constituents. Similar to the use of nanoscale catalysts in the manufacturing process, the use of nanomaterials to treat process waste streams (gas, liquid, or solid phases) provides enhancements in removal rates and/or efficiencies. One key activity will include the application of nano catalysis for the reduction of air pollutants and a better understanding of how these catalysts can be used in various environmental applications. Inorganic nanoscale materials, including metallic iron nanomaterials and aluminosilicate-based zeolites, have been synthesized for removal or degradation of metals and organic contaminants from air and water effluents generated as a result of manufacturing and power-generation operations (Ponder 2000, Song 2005). Similar to the case described above for the manufacturing process, the use of nanomaterials in end-of-pipe treatments affords the opportunity for regeneration or controlled disposal of treatment by-products. In addition, this research will study the use of nano-scale iron particles to remediate aqueous streams contaminated with chlorinated-organics, pesticides, PCBs, heavy metals and such inorganics like Cr+6, arsenates, perchlorates, and nitrates. If these treatment and remediation processes are successful, they can be incorporated into existing treatment systems to further reduce contaminant loading.

Another area of emphasis within this program will be to investigate the ability to physically and chemically tailor substances, surfaces, and pores at the nano-scale to improve selectivity and efficiency of membrane filtration, adsorption, and catalysis. The objective is to identify and evaluate innovative, high performance or lower cost alternatives for treating critical contaminants. Improvements for many different treatment scenarios (e.g., matrices, contaminants, treatment technologies, and treatment goals) may become feasible. Examples of areas where such an approach could provide significant improvements in removal performance and cost savings is the use of nanotechnology to produce advanced sorbents for mercury control and water treatment. In the mercury area, the ability to directly link the physical and chemical nature of binding sites in the materials with the performance of those materials is the key to developing new or improved adsorbents with properties that exceed those that have conventionally been used. In the water area, nanomaterials may enable the manufacture of media that are more selective, efficient, and economical for removal or destruction of existing or emerging contaminants from drinking water, wastewater, and storm water. These improved media may arise from better design and uniformity of pore size, particle size, or composition made feasible by nano-scale design and control of the manufacturing process.

Remediation of contaminated sites is another area where ORD will explore the use of nanomaterials. Examples of these research and development efforts include the development of nanoscale metallic solids or biopolymers for the destruction of organic contaminants or the extraction of inorganic contaminants from ground water and soil. Ultimately, EPA can play a significant role in advancing the development and implementation of these technologies through research and testing. Using past experience implementing waste minimization, treatment, and remediation technologies, EPA can fulfill the much-needed role of a technical mediator between the commercial entities actively pursuing development of synthetic nanomaterials and those who may be negatively affected by the large-scale utilization of these materials.

4.4.4. Anticipated Outcomes

Within this research theme, the near-term emphasis will be on addressing scientific questions related to the first two outcomes listed below.

- An evaluation of the efficacy of existing pollution control approaches and technologies to manage releases of engineered nanomaterials to all media during their production
- The results of this assessment will be provided in the form of reports and computer- based systems that can be used to identify and address the unique issues associated with various industrial operations. Ultimately, regulatory officials and industry could be informed about whether there are potential risks posed by the releases of engineered nanomaterials into the environment and what potential controls might be available to limit potential risks. This has potential to influence decisions regarding manufacturing, importing, storing, handling, and using of selected nanomaterials.
- ORD will collaborate with others to report on opportunities to reduce the environmental implications of nanomaterial production by employing greener synthesis approaches.

- ORD will identify design production processes that minimize or eliminate any emissions/releases and reduce energy consumption during the manufacturing of nanomaterials and products.
- ORD will report on the viability, performance, and benefits of the use of nanotechnology for the abatement and remediation of conventional toxic pollution.

5.0. Implementation, Research Linkages, and Communication

Implementation

The research described in this NRS will be implemented through the multi-year plan (MYP) process. ORD uses MYPs to provide a link between the strategic plans and annual plans, showing how we intend to meet our out year goals. The MYPs chart the direction of ORD's research program in selected topic areas over a period of approximately five to ten years. The MYPs also link to each other, showing how the different parts of ORD's research areas are integrated. MYPs aid in the evaluation of research options and foster the integration of strategic risk-based environmental protection and anticipation of future environmental issues. They also allow for a more comprehensive understanding of any changes needed to emphasize a new direction or accelerate an existing program. MYPs are updated periodically to reflect changes in Agency strategic thinking, the realities of available resources, and the current state-of-the-science.

ORD has formed a Nanomaterial Research Coordination Team, which is a cross-Agency research planning group, to communicate program office and regional research needs to ORD and for ORD to communicate its research activities and products under the strategic research themes. This approach promotes ORD's focus on the highest priority issues and provides a roadmap to achieving our long-term research goals while allowing the flexibility for ORD to address emerging nanotechnology issues that are affecting specific programmatic areas.

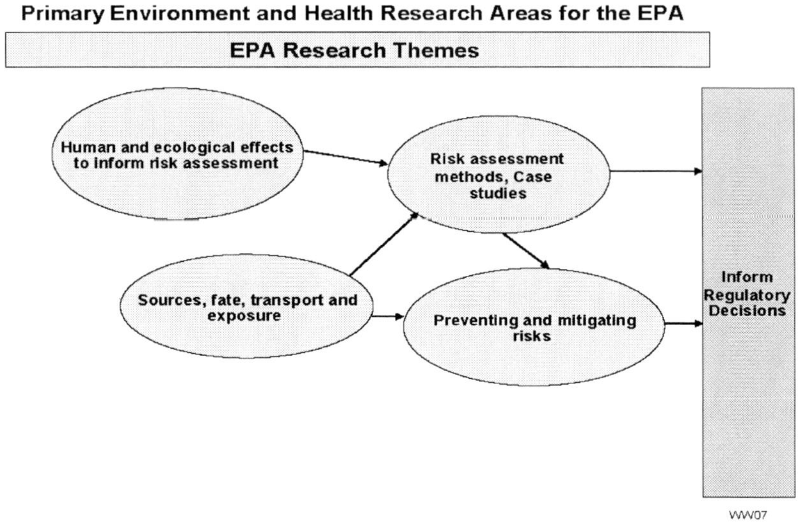

Figure 5-1. Environmental and Health Research Theme Linkages.

Selection of Primary Engineered Nanomaterials – Initial Focus for Study

The ORD NRS Team has decided to focus on five engineered nanomaterials for study. The materials selected are: (1) titanium dioxide; (2) zero valent iron; (3) nanosilver; (4) carbon nanotubes; and (5) cerium oxide. These materials were selected with the goal of developing predictive models and tools that will enable representative classes of nanomaterials to be tested in lieu of individual materials.

Linkages to Related Federal Research

Figure 5-1 displays the flow of the EPA research themes to support each other and to inform decisions. EPA will rely on basic research conducted by other Federal agencies to support EPA applied research. NSF and NIEHS will contribute much of the basic research on biomedical, engineering, and material development and characterization. ORD and NTP scientists are working to prioritize/evaluate toxicity testing and developing approaches to predict toxicity, while NIST will provide nanomaterial characterization and analytical standards to provide a common context for the Federal research programs. ORD's research program is coordinated and leveraged with the other Federal agencies involved in nanotechnology environment, health, and safety research through various collaborative activities. For example, NIOSH and NTP collaboration on the toxicology of carbon nanotubes and will support ORD health effects research and assessment.

6.0. REFERENCES

Davis, J. M. (2007). How to assess the risks of nanotechnology: learning from past experience. *J. Nanosci. Nanotechnol. 7(2),* 402-409.

Davis, J. M. & Thomas, V. M. (2006). Systematic approach to evaluating trade-offs among fuel options: the lessons of MTBE, Ann. N.Y. *Acad. Sci. 1076*, 498-515.

Morgan, K. (2005). Development of a Preliminary Framework for Informing the Risk Analysis and Risk Management of Nanoparticles. *RiskAnalysis 25*, No. 6, 162 1-1635.

Dix et al., *Toxicol Sci., 95(1),* 5-12, 2007

Environmental Defense - DuPont Nano Partnership (2007) Nano risk framework. New York, NY: Environmental Defense. Available at http://www.environmentaldefense.org/go/nano.

Maynard, A.D. (2006). *Nanotechnology: A research strategy for addressing risk.* Woodrow Wilson International Center for Scholars. PEN 3 July. Washington, D.C.

Morgan, K. (2005). Development of a Preliminary Framework for Informing the Risk Analysis and Risk Management of Nanoparticles. *RiskAnalysis 25*, No. 6, 162 1-1635.

National Nanotechnology Initiative, Sept.2006. (www.nano.gov/NNI_EHS_research_needs.pdf)

National Research Council of the National Academy of Sciences (www.nap.edu/catalog/1 1970.html#toc)

National Research Council (NRC, 1983). *RiskAssessment in the Federal Government: Managing the Process.* National Academy Press, Washington, DC.

Nowack, B. & Bucheli, T. D. (2007). Occurrence, behavior and effects of nanoparticles in the environment. *Environmental Pollution.* In Press (Corrected proof available online).

Nishioka, Y., Levy, J. I., Norris, G. A., *et al.* (2002). Integrating risk assessment and life cycle assessment: a case study of insulation. *RiskAnalysis, 22*, 1003-1017.

Ponder, S. M., Darab, J. G. & Mallouk, T. E. (2000). Remediation of Cr(VI) and Pb(II) aqueous solutions using supported, nanoscale zero-valent iron. *Environmental Science and Technology, 34*, 2564 -2569.

Presidential/Congressional Commission on Risk Assessment and Risk Management (1997). *Framework for Environmental Health Risk Management.* Final Report of the Commission. Volume 1

Science Policy Council (SPC, 2007). U.S. Environmental Protection Agency Nanotechnology White Paper EPA 100/B-07/001, February 2007

Schmidt, Karen F., Green Nanotechnology: It's easier than you think. Woodrow Wilson International Center for Scholars, April 2007

Shatkin, J. A. & Qian, A. (2004). Classification schemes for priority setting and decision making: a selected review of expert judgment, rule-based, and prototype methods. In *Comparative Risk Assessment and Environmental Decision Making. Linkov, I. & A. Ramadan, Eds.: 2* 13-244, Luewer, Amsterdam.

Song, W., Li, G., Grassian, V. H. & Larsen, S. C. (2005). Development of improved materials for environmental applications: Nanocrystalline NaY zeolites. *Environmental Science and Technology, 39,* 1214-1220.

Sonneman, G., Castells, F. & Schumacher, M. (2004) *Integrated Life-Cycle and Risk Assessment for Industrial Processes.* Lewis Publishers. Boca Raton, FL.

Surowiecki, J. (2004). *The Wisdom of Crowds.* Little Brown, London

U.S. Environmental Protection Agency (1998). Guidelines for ecological risk assessment. Washington, DC: Office of Research and Development, U.S. Environmental Protection Agency. EPA/630/R-95/002F.

U.S. Environmental Protection Agency (2007). U.S. Environmental Protection Agency Nanotechnology White Paper. Washington, DC: Science Policy Council, U.S. Environmental Protection Agency. EPA 100/B-07/001.

APPENDIX A. RELATIONSHIP OF ORD RESEARCH STRATEGY TO EPA WHITE PAPER RESEARCH NEEDS (CURRENT RESEARCH (CR), SHORT-TERM RESEARCH (SR), AND LONG-TERM RESEARCH (LT))

The table below is only intended to link the activities described in this strategy with the overall research need questions in the EPA White Paper. It is not designed to provide details on implementation of the NRS.

Research Theme	Research Need Questions (from EPA White Paper; EPA, 2007)	Relationship to ORD Strategy (CR, SR, LR)*
Research Needs for Risk Assessment		
Chemical Identification and Characterization	What are the unique chemical and physical characteristics of nanomaterials? How do these characteristics vary among different classes of materials (e.g., carbon based, metal based) and among the individual members of a class (e.g., fullerenes, nanotubes)?	SR
	How do these properties affect the material's reactivity, toxicity and other attributes?	SR
	To what extent will it be necessary to tailor research protocols to the specific type and use pattern of each nanomaterial? Can properties and effects be extrapolated within class of nanomaterials?	SR
	Are there adequate measurement methods/technology available to fully characterize nanomaterials, to distinguish among different types of nanomaterials, and distinguish	SR

(Continued)

Research Theme	Research Need Questions (from EPA White Paper; EPA, 2007)	Relationship to ORD Strategy (CR, SR, LR)*
	intentionally produced nanomaterials from ultrafine particles or naturally occurring nanosized particles?	
	Are current test methods for characterizing nanomaterials adequate for the evaluation hazard and exposure data?	SR
	Do nanomaterial characteristics vary from their pure form in the laboratory to their form as components of products and eventually to the form in which they occur in the environment?	SR
	What intentionally produced nanomaterials are now on the market and what new types of materials can be expected to be developed?	SR
	How will manufacturing processes, formulations, and incorporations in end products alter the characteristics of nanomaterials?	SR
Environmental Fate and Treatment Research Needs		
Transport Research Questions	What are the physical and chemical factors that influence the transport and deposition of intentionally produced nanomaterials in the environment? How do nanomaterials move through these media? Can existing information on soil colloidal fate and transport and atmospheric ultrafine	SR

Research Theme	Research Need Questions (from EPA White Paper; EPA, 2007)	Relationship to ORD Strategy (CR, SR, LR)*
	particulate fate and transport inform our thinking?	
	How are nanomaterials transported in the atmosphere? What nanomaterials properties and atmospheric conditions control the atmospheric fate of nanomaterials?	SR
	To what extent are nanomaterials mobile in soils and in groundwater? What is the potential for these materials, if released to soil or landfills, to migrate to groundwater and within aquifers, with potential exposure general populations via groundwater ingestion?	CR
	What is the potential for these materials to be transported bound to particulate matter, sediments, or sludge in surface waters?	SR
	How do the aggregation, sorption and agglomeration of nanoparticles affect their transport?	SR
	How do nanomaterials bioaccumulate? Do their unique characteristics affect their bioavailability? Do nanomaterials bioaccumulate to a greater or lesser extent than macro-scale or bulk materials?	SR
Transformation Research Questions	How do nanoparticles react differently in the environment than their bulk counterparts	SR

(Continued)

Research Theme	Research Need Questions (from EPA White Paper; EPA, 2007)	Relationship to ORD Strategy (CR, SR, LR)*
	What are the physical and chemical factors that impact the persistence of intentionally produced nanomaterials in the environment? What data are available on the physical and chemical factors that affect the persistence to unintentionally produced nanomaterials (e.g., carbon-based combustion products) that may provide information regarding intentionally produced nanomaterials?	SR
	Do particular nanomaterials persist in the environment, or undergo degradation via biotic of abiotic processes? If they degrade, what are the byproducts and their characteristics? Is the nanomaterial likely to be in the environment, and thus be available for bioaccumulation/biomagnification?	SR
	How are the physical, chem.-ical and biologic properties of nanomaterials altered in complex environmental media such as air, water, and soil? How do redox processes influence environmental transformation of nanomaterials? To what extent are nanomaterials photoreactive in the atmosphere, in water, or on environmental surfaces?	SR

Research Theme	Research Need Questions (from EPA White Paper; EPA, 2007)	Relationship to ORD Strategy (CR, SR, LR)*
	How do the aggregation, sorption and agglomeration of nanoparticles affect their transport?	SR
	In what amounts and in what forms may nanoparticles be released from materials that contain them, as a result of environment forces (rain, sunlight, etc.) or through use, re-use, and recycle or disposal.	LR
Chemical Interaction Research Questions	How do nanosized adsorbents and chemicals sorbed to them in influence their respective environmental interactions? Can these materials alter the mobility of other substances in the environment? Can these materials alter the reactivity of other substances in the environmental?	SR
Treatment Research Questions	What is the potential for these materials to bind to soil, subsurface materials, sediment or wastewater sludge, or binding agents in waste treatment facilities?	SR
	Are these materials effectively removed from wastewater using conventional wastewater treatment methods and, if so, by what mechanism?	SR
	Do these materials have an impact on the treatability of other substances in waste streams (e.g., wastewater,	SR

(Continued)

Research Theme	Research Need Questions (from EPA White Paper; EPA, 2007)	Relationship to ORD Strategy (CR, SR, LR)*
	hazardous and nonhazardous solid wastes), or on treatment facilities performance?	
	How effective are existing treatment methods (e.g., carbon adsorption, filtration, coagulation and settling, or incineration/air pollution control system sequestration/ stabilization) for treating nanomaterials?	SR
Assessment Approaches and Tools Questions	Can existing information on soil colloidal fate and transport, as well as atmospheric ultrafine particulate fate and transport, inform our thinking? Do the current databases of ultrafines/fibers shed light on any of these questions?	CR
	Do the different nanomaterials act similarly enough to be able to create classes of like compounds? Can these classes be used to predict structure-activity relationships for future materials?	CR
	Should current fate and transport models be modified to incorporate the unique characteristics of nanomaterials?	SR
Environmental Detection an Analysis Research Needs		
Existing Methods and Technologies Research Questions	Are existing methods and technologies capable of detecting, characterizing, and	SR

(Continued)

Research Theme	Research Need Questions (from EPA White Paper; EPA, 2007)	Relationship to ORD Strategy (CR, SR, LR)*
	quantifying intentionally produced nanomaterials by measuring particle number size, shape, surface properties (e.g., reactivity, charge, and area), etc? Can they disting-uish between intentionally produced nanomaterials ofinterest and other ultrafine particles? Can they distinguish between individual particles of interest and particles that may have agglomerated or attached to larger particles?	
	Are standard procedures available for both sample preparation and analysis?	SR
	Are quality assurance and control reference materials and procedures available?	SR
	How would nanomaterials in waste media be measured and evaluated?	SR
New Methods and Technologies Research Need	What low-cost, portable, and easy-to-use technologies can detect, characterize, and quantify nanomaterials of interest in environmental media and for personal exposure	LR
Human Exposures, Their Measurement and Control		
Risk and Exposure Assessment Research Questions	Is the current exposure assessment process adequate for assessing exposures to nanomaterials? Is mass dose an effective metric for	SR

(Continued)

Research Theme	Research Need Questions (from EPA White Paper; EPA, 2007)	Relationship to ORD Strategy (CR, SR, LR)*
	measuring exposure? What alternative metric (e.g., particle count, surface area) should be used to measure exposure? Are sensitive populations' (e.g., endangered species, children, asthmatics, etc.) exposure patterns included?	
	How do physical and chemical properties of nanomaterials affect releases and exposures?	SR
	How do variations in manufacturing and subsequent processing, and the use of particle surface modifications affect exposure characteristics?	SR
Release and Exposure Quantification Research Questions	What information is available about unique release and exposure patterns of nanomaterials? What additional information needed?	LR
	What tools/resources currently exist for assessing releases and exposures within EPA (chemical release information/ monitoring systems (e. g., TRI), measurement tools, models, etc)? Are these tools/ resources adequate to measure, estimate, and assess releases and exposures to nanomaterials? Is degradation of nanomaterials accounted for?	SR

Research Theme	Research Need Questions (from EPA White Paper; EPA, 2007)	Relationship to ORD Strategy (CR, SR, LR)*
	What research is needed to develop sensors that can detect nanomaterials, including personal exposure monitoring?	SR
Release and Exposure Reduction and Mitigation Research Questions	What tools/resources exist for limiting release and/or exposure during manufacture, use of following release via waste streams? Are these tools/ resources adequate for nanomaterials?	SR
	Are current respirators, filters, gloves, and other PPE capable of reducing or eliminating exposure from nanomaterials?	LR
	Are current engineering controls and polluti prevention devices capable of minimizing releases and exposures to nanomaterials?	SR
	Are technologies and procedures for controlling spills during manufacture and use adequate for nanomaterials? Can current conventional technologies (i.e., for non-nanomaterials) be adapted to control nanomaterial spills?	LR
	In the case of an unintentional spill, what are the appropriate emergency actions? How are wastes from the response actions disposed of properly?	LR
	Do existing methods using vacuum cleaners with HEPA filters work to clean up spill of solid nanomaterials? If not,	LR

(Continued)

Research Theme	Research Need Questions (from EPA White Paper; EPA, 2007)	Relationship to ORD Strategy (CR, SR, LR)*
	would a wet vacuum system work?	
	What PPEs would be suitable for use by operators during spill mitigation?	LR
Human Health Effects Assessment Research Needs	What are the health effects (local and systemic; acute and chronic) from either direct exposure to nanomaterials, or to their byproducts, associated with those nanotechnology applications that are most likely to have potential for exposure?	CR
	Are there specific toxicological endpoints that are of higher concern for nanomaterials, such as neurological, cardiovascular, respiratory, or immunological effects, etc.?	SR
	Are current testing methods (organisms, exposure regimes, media, analytical methods, testing schemes) applicable to testing nanomaterials in standardized agency toxicity tests (http://www.epa.gov/opptsfrs/OPPTS_Harmonized/)?	SR
	Are current test methods, for example OECD and EPA harmonized test guidelines, capable of determining the toxicity of the wide variety of intentionally produced nanomaterials and byproducts associated with their production and applications?	SR

Research Theme	Research Need Questions (from EPA White Paper; EPA, 2007)	Relationship to ORD Strategy (CR, SR, LR)*
	Are current analytical methods capable of analyzing and quantifying intentionally produced nanomaterials to generate dose- response relationships?	SR
	What physical and chemical properties regulate nanomaterial absorption, distribution, metabolism, and excretion (ADME)?	SR
	What physical and chemical properties and dose metrics best correlate with the toxicity (local and systemic; acute and chronic) of intentionally produced nanomaterials following various routes of exposure?	CR
	How do variations in manufa-cturing and subsequent proce-ssing, and the use of particle surface modifications affect nanomaterial hazard?	CR
	Are there subpopulations that may be at increased risk of adverse health effects associ-ated with exposure to intentio-nally produced nanomaterials?	SR
	What are the best approaches to build effecti predictive models of toxicity (SAR, PBPK, "omics", etc.)?	SR
	Are there approaches to grou-ping particles in classes relat-ive to their toxicity potencies, in a manner that links in vitro, in vivo, and in silico data?	LR

(Continued)

Research Theme	Research Need Questions (from EPA White Paper; EPA, 2007)	Relationship to ORD Strategy (CR, SR, LR)*
Ecological Effects Research Needs	Are current testing schemes and methods (organisms, endpoints, exposure regimes, media, analytical methods) applicable to testing nanomaterials in standardized toxicity tests? Both pilot testing protocols and definitive protocols should be evaluated with respect to their applicability to nanomaterials.	SR
	What is the distribution of nanomaterials in ecosystems? Research on model ecosystems studies (micro, mesocosms) is needed to assist in determining the distribution of nanomaterials in ecosystems and potentially affected compartments and species.	SR
	What are the effects (local and systemic; acu and chronic) from either direct exposure to nanomaterials, or to their byproducts, associated with those nanotechnology applications that are most likely to have potential for exposure?	SR
	What are the absorption, distribution, metabolism, elimination (ADME) parameters for various nanomaterials for ecological receptors? This topic addresses the uptake, transport, bioaccumulation relevant to a range of species	SR

Research Theme	Research Need Questions (from EPA White Paper; EPA, 2007)	Relationship to ORD Strategy (CR, SR, LR)*
	(fish, invertebrates, birds, amphibians, reptiles, plants, microbes).	
	How do variations in manufacturing and subsequent processing, and the use of particl surface modifications affect nanomaterial toxicity to ecological species?	SR
	What research is needed to examine the interaction of nanomaterials with microbes in sewage treatment plants, in sewage effluent, and in natural communities of microbes in natural soil and natural water?	LR
	What research is needed to develop structure activity relationships (SARs) for nanomaterials for aquatic organisms?	SR
	What are the modes of action (MOAs) for various nanomaterials for ecological species Are the MOAs different or similar across ecological species?	SR
Risk Assessment Research - Case Study	Which of the research needs identified in the EPA Nanotechnology White Paper and in the overarching and component questions listed here are of the highest priority from the standpoint of generating information needed to support risk assessments of nanomaterials selected as case studies?	SR

(Continued)

Research Theme	Research Need Questions (from EPA White Paper; EPA, 2007)	Relationship to ORD Strategy (CR, SR, LR)*
	For selected case studies, using expert judgment methods, what do we know and what do we need to know (in priority ranking) regarding the potential for exposure (cumulative and aggregate) of humans and biota to primary and secondary materials via multi-media pathways?	SR
	Which nanomaterials and applications should ORD focus its efforts on first as case studies? Which expert judgment method(s) is (are) applicable to evaluating selected case studies for identifying "what we know and what we need to know" and for prioritizing research needs?	SR
	For selected case studies, using expert judgment methods, what do we know and what do we need to know (in priority ranking) regarding specific details of product life cycle stages, including feedstocks, manufacturing, distribution, storage, use, and disposal/reuse?	SR
	For selected case studies, using expert judgment methods, what do we know and what do we need to know (in priority ranking) regarding likely primary nanomaterials and secondary substances (e.g., waste by-products) that	SR

Research Theme	Research Need Questions (from EPA White Paper; EPA, 2007)	Relationship to ORD Strategy (CR, SR, LR)*
	may be released/emitted at each stage of the product life cycle?	
	For selected case studies, using expert judgment methods, what do we know and what do we need to know (in priority ranking) regarding likely environmental media (air, water, soil, food web) to which releases/ emissions of primary and secondary materials may occur, and about potential transport and fate processes that may be applicable?	SR
Green Manufacturing Research Needs	How can nanotechnology be used to reduce waste products during manufacturing?	SR
	How can nanomaterials be made using benign starting materials?	CR
	How can nanotechnology be used to reduce the resources needed for manufacturing (both materials and energy)?	LR
	What is the life cycle of various types of nanomaterials and nanoproducts under a variety of manufacturing and environmental conditions?	SR
Green Energy Research Needs	What research is needed for incentives to encourage nanotechnology to enable green energy?	LR
	How can nanotechnology assist "green" energy production, distribution, and use?	LR

(Continued)

Research Theme	Research Need Questions (from EPA White Paper; EPA, 2007)	Relationship to ORD Strategy (CR, SR, LR)*
Environmental Remediation/Treatment Research Needs	Which nanomaterials are most effective for remediation and treatment?	CR
	What are the fate and effects of nanomateria used in remediation applications? When nanomaterials are placed in groundwater treatment, how do they behave over time?	CR
	Do they move in groundwater? What is their potential for migrating to drinking water wells?	CR
	How can we improve methods for detecting and monitoring nanomaterials used in remediation and treatment?	SR
	To what extent are these materials and their byproducts persistent, bioaccumulative, and toxic and what organisms are affected?	SR
	If toxic byproducts are produced, how can these be reduced?	LR
	What is needed to enhance the efficiency an cost-effectiveness of remediation and treatment technology?	SR
Sensors	How can nanomaterials be employed in the development of sensors to detect biological and chemical contaminants?	LR
	How can systems be developed to monitor agents in real time and the resulting data accessed remotely?	LR

Research Theme	Research Need Questions (from EPA White Paper; EPA, 2007)	Relationship to ORD Strategy (CR, SR, LR)*
	How these small-scale monitoring systems b developed to detect personal exposures and in vivo distributions of toxicants.	LR

Appendix B. Description of EPA Office of Research and Development

The Office of Research and Development (ORD) is the principal research arm of the Environmental Protection Agency (EPA) (*http://www.epa.gov /ord/)*. Its role is to provide the critical science for the Agency's environmental decision-making. Unlike much of EPA, ORD has no direct regulatory function; its responsibility is to inform the policymaking process. Through the development of technical information and scientific tools, ORD's research strengthens EPA's science base by providing its program offices and regional offices with sound scientific advice and information for use in developing and implementing scientifically defensible environmental policies, regulations, and practices.

As may be seen in Figure B-1, ORD is led by the Assistant Administrator (AA) for Research and Development, who reports directly to the EPA Administrator. This position involves providing leadership in establishing research priorities, ensuring the means for technical evaluation and peer-review of ORD's products, and contributing scientific input into the EPA's regulatory decisions.

The AA ORD is supported by a Deputy Assistant Administrator (DAA) for Management and a DAA for Science. The Directors of ORD's Laboratories Centers provide scientific leadership relative to their respective organizations and report to the AA ORD. Recently, ORD established National Program Directors (NPDs). The NPDs provide a strategic vision of the stakeholder needs and overall coordination of research programs delineated in ORD Multi-Year Plans (MYPs).

ORD is comprised of seven national Laboratories and Centers and two Offices. The Laboratories and Centers, spread across the country, conduct research across the risk assessment/risk management paradigm related to both the environment and human health. ORD also has a National Homeland Security Research Center and a National Center for Computational Toxicology. ORD's two offices are the Office of Science Policy (OSP) and the Office of Resources Management and Administration (ORMA). OSP plays a vital role by providing expert advice and evaluation on the use of scientific knowledge and science policy to support sound science in the Agency. OSP accomplishes this mission by leading efforts in science integration, coordination and communication across ORD, and between ORD and the Agency's programs, regions, and external parties. ORMA manages a broad spectrum of issues and provides counsel/advice on all matters relating to the responsible management of ORD's resources.

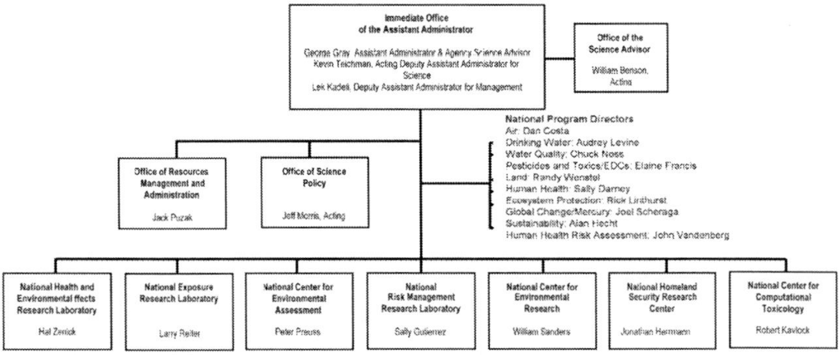

Figure B-1. Organization Chart for the Office of Research and Development.

End Notes

[1] Use of the term nanomaterials refers to engineered nanomaterials and particles.
[2] A brief description of EPA Office of Research and Development is presented in Appendix B
[3] A raw material required for an industrial process

In: Nanomaterial Research Strategy
Editors: Earl B. Purcell
ISBN: 978-1-60876-845-5

Chapter 2

NANOTECHNOLOGY AND ENVIRONMENTAL, HEALTH, AND SAFETY: ISSUES FOR CONSIDERATION

John F. Sargent

SUMMARY

Nanotechnology—a term encompassing nanoscale science, engineering, and technology—is focused on understanding, controlling, and exploiting the unique properties of matter that can emerge at scales of one to 100 nanometers. A key issue before Congress regarding nanotechnology is how best to protect human health, safety, and the environment as nanoscale materials and products are researched, developed, manufactured, used, and discarded. While the rapidly emerging field of nanotechnology is believed by many to offer significant economic and societal benefits, some research results have raised concerns about the potential adverse environmental, health, and safety (EHS) implications of nanoscale materials.

Some have described nanotechnology as a two-edged sword. On the one hand, some are concerned that nanoscale particles may enter and accumulate in vital organs, such as the lungs and brains, potentially causing harm or death to humans and animals, and that the diffusion of nanoscale particles in the environment might harm ecosystems. On the other hand, some believe that nanotechnology has the potential to deliver important EHS benefits such as reducing energy consumption, pollution, and greenhouse gas emissions; remediating environmental damage; curing, managing, or preventing diseases; and offering new safety-enhancing materials that are stronger, self-repairing, and able to adapt to provide protection.

Stakeholders generally agree that concerns about potential detrimental effects of nanoscale materials and devices—both real and perceived—must be addressed to protect and improve human health, safety, and the environment; enable accurate and efficient risk assessment, risk management, and cost-benefit trade-offs; foster innovation and public confidence; and ensure that society can enjoy the widespread economic and societal benefits that nanotechnology may offer. Congressionally-mandated reviews of the National Nanotechnology Initiative (NNI) by the National Research Council and the President's Council of Advisors on Science and Technology have concluded that additional research is required to make a rigorous risk assessment of nanoscale materials.

INTRODUCTION

Nanotechnology—a term encompassing nanoscale science, engineering, and technology—is focused on understanding, controlling, and exploiting the unique properties that can emerge at scales of one to 100 nanometers.[1] These properties are believed by many to offer substantial economic and societal benefits.

A key issue before Congress regarding nanotechnology is how best to protect human health, safety, and the environment as nanoscale materials and products are researched, developed, manufactured, used, and discarded. While the rapidly emerging field of nanotechnology is believed by many to offer significant economic and societal benefits, some research results have raised concerns about the potential environmental, health, and safety (EHS) implications of nanoscale materials. Potential tools the Federal government might use to address these concerns include research and development, regulation, and international engagement.

Some of the properties of nanoscale materials (e.g., small size, high surface area-to-volume ratio) that have given rise to great hopes for beneficial applications have also given rise to concerns about their potential adverse implications for the environment, and human health and safety.[2] There are more than 600 nanotechnology products reportedly commercially available,[3] and with this number of products concerns have been raised about the health and safety of the scientists working with nanoscale materials, workers who manufacture the products, consumers who use the products, and members of the general public who may be exposed to nanoparticles, as well as the environmental impact of nanomanufacturing processes and the use and disposal of nanotechnology products.

Nanoscale particles can result from a variety of different processes. While nanoscale particles can occur naturally (e.g., some particles produced by forest fires, sea spray, volcanoes) and as an incidental by-product of human activities (e.g., some particles contained in welding fumes, diesel exhaust, industrial effluents, cooking smoke), EHS concerns have focused primarily on nanoscale materials that are intentionally designed and produced, often referred to as engineered nanomaterials.

Issues surrounding the potential EHS implications of nanotechnology emerged with the launch in 2000 of the National Nanotechnology Initiative (NNI). The NNI is a multi-agency federal effort to coordinate and expand federal nanotechnology research and development (R&D) efforts. Between FY2001 and FY2009, the federal government invested \$9.9 billion in nanotechnology R&D, including approximately \$1.5 billion in FY2009. Many governments around the world have followed the U.S. lead and established their own national nanotechnology programs. The private sector has invested heavily as well. Global nanotechnology R&D investments—public and private—are estimated to have totaled \$12.4 billion in 2006 alone.[4]

Such large investments and intensified efforts to capitalize on these public and private investments have caused some observers (as detailed later in this report) to suggest that there is insufficient information about the potential effects nanotechnology products and manufacturing processes may have on human health, safety, and the environment. They assert a variety of uncertainties, including: how nanoscale particles might be transported in air, water, and soil; how they might react with the environment chemically, biologically, or through other processes; how they might be distributed and deposited; and whether they might accumulate in plants or animals.

Others express the view that concerns about nanotechnology EHS implications are often overgeneralized and overstated. Among the arguments they put forth are that nanoscale materials are frequently embedded in other materials as part of the manufacturing process; that some nanotechnology products, such as semiconductors, have nanoscale features but do not contain nanoscale particles; that nanotechnology materials may replace other materials that have significant and known risks; that some nanoscale particles tend to aggregate or agglomerate in the environment into larger particles that no longer have nanoscale dimensions; and that people are regularly exposed to nanoscale particles produced naturally and as incidental by-products of human activities.

Congressionally-mandated reviews of the NNI by the National Research Council (NRC) and the President's Council of Advisors on Science and Technology (PCAST) have concluded that additional rcscarch is required to make a rigorous risk assessment of nanoscale materials. In addition, the NRC warned that, until such information is available, precautionary measures should be taken to protect the health and safety of workers, the public, and the environment.

Nevertheless, most stakeholders agree that these concerns about the potential detrimental effects of nanoscale materials and devices—both real and perceived—must be addressed. Among the issues these stakeholders have identified are characterizing the toxicity of nanoscale materials; developing methods for assessing and managing the risks of these materials; and understanding how materials move in, and interact with, the environment.

This report identifies the potential environmental, health, and safety opportunities and challenges of nanotechnology; explains the importance of addressing nanotechnology EHS concerns; identifies and discusses nanotechnology EHS issues; and summarizes options for Congressional action, including the nanotechnology EHS-related provisions of selected legislation. The report also includes two appendices. Appendix A provides an overview of selected nanotechnology EHS activities of federal regulatory agencies. Appendix B provides an overview of selected EHS-related international engagement efforts of NNI agencies.

Opportunities and Challenges

Historically, many new technologies have delivered general societal benefits while presenting EHS challenges. For example, automobiles increased personal mobility and provided faster, less expensive transportation of goods, but soon became a leading cause of accidental deaths and injuries, as well as a source of emissions that can damage air quality and may contribute to global climate change. Similarly, genetically-modified (GM) plants have traits such as greater resistance to pests, pesticides, or cold temperatures that contribute to higher crop yields, while critics argue some GM foods contribute to food allergies and antibiotic resistance.[5]

Like other new technologies, nanotechnology offers potential economic and societal benefits, and presents potential EHS challenges as well. Nanotechnology advocates assert, however, that nanotechnology provides the opportunity to reduce or eliminate known risks by engineering around them. Proponents maintain that nanotechnology also offers the potential for significant EHS benefits, including:

- reducing energy consumption, pollution, and greenhouse gas emissions;
- cleaner, more efficient industrial processes;
- remediating environmental damage;
- curing, managing, or preventing deadly diseases; and
- offering new materials that protect against impacts, self-repair to prevent catastrophic failure, or change in ways that protect or aid soldiers on the battlefield.

For example, nanoscale materials show promise for detecting, preventing, and removing pollutants. According to the Environmental Protection Agency (EPA):

> nanoscale cerium oxide has been developed to decrease diesel engine emissions; iron nanoparticles can remove contaminants from soil and ground water; and nano-sized sensors hold promise for improved detection and tracking of contaminants.[6]

In the area of human health, scientists assert nanotechnology has the potential for improving disease diagnostics, sensing, monitoring, assessment, and treatment. In particular, the National Cancer Institute (NCI) views nanotechnology as likely to provide revolutionary tools to extend and improve lives. In July 2004, NCI launched a five-year, $145 million initiative focused on applying nanotechnology to the prevention, detection, and treatment of cancer and amelioration of its symptoms. At the initiative's launch, then-NCI Director Andrew von Eschenbach identified nanotechnology as a key component of the agency's strategy for ending death and suffering from cancer by 2015 (see text box, "Potential Nanotechnology Cancer Applications").[7]

Some characteristics of nanoscale particles could produce both positive and negative consequences. According to E. Clayton Teague, director of the National Nanotechnology Coordination Office (NNCO),

> the unique properties of these [nanotechnology] materials are a double-edged sword: they can be tailored for beneficial properties, but also have unknown consequences, such as new toxicological and environmental effects.[8]

POTENTIAL NANOTECHNOLOGY CANCER APPLICATIONS

The NCI Cancer Nanotechnology Plan asserts that nanotechnology can serve as an enabling technology for a variety of cancer-related applications:

- imaging agents and diagnostics that allow clinicians to detect cancer in its earliest, most easily treatable, pre-symptomatic stage;
- systems that provide real-time assessments of therapeutic and surgical efficacy;
- multifunctional, targeted devices capable of bypassing biological barriers to deliver therapeutic agents at high local concentrations directly to cancer cells and tissues that play a critical role in the growth and metastasis of cancer;
- agents capable of monitoring predictive molecular changes and preventing precancerous cells from becoming malignant;
- surveillance systems that detect mutations that may trigger the cancer process and genetic markers that indicate a predisposition for cancer;
- novel methods for managing the symptoms of cancer that adversely impact quality of life; and
- research tools that enable investigators to quickly identify new targets for clinical development and predict drug resistance.

Source: *Cancer Nanotechnology Plan: A Strategic Initiative to Transform Clinical Oncology and Basic Research Through the Directed Application of Nanotechnology*, National Cancer Institute, National Institutes of Health, Department of Health and Human Services, July 2004.

The following examples illustrate how the same nanotechnology material may be both potentially beneficial and potentially harmful:

- Nanoscale silver is highly effective as an antibacterial agent in wound dressings, clothing, and washing machines, but some have expressed concerns that widespread dispersion of nanoscale silver in the environment could kill microbes that are vital to waste water treatment plants and to ecosystems. Some beneficial bacteria, for example, break down organic matter, remove nitrogen from water, aid in animal digestion, protect against fungal infestations, and even aid some animals in defense against predators.[9]
- Some nanoscale particles may have the potential to penetrate the blood-brain barrier, a structure that protects the brain from harmful substances in the blood but also hinders the delivery of therapeutic agents. The characteristics of certain nanoscale materials may allow pharmaceuticals to be developed to purposefully and beneficially cross this barrier and deliver medicine directly to the brain to treat, for example, a brain tumor.[10] Some critics are concerned, however, that nanoscale particles might unintentionally pass through the blood-brain barrier causing harm to humans and animals.[11]
- Certain nanoscale materials are highly chemically reactive due to their high surface-to-volume ratio.[12] This is a property that might be positively exploited in catalysis, treatment of groundwater contamination, and site remediation. This property also is being explored for use in protective masks and clothing as a defense against chemical and biological agents. However, some research results indicate that the reactivity of some nanoparticles potentially can result in cell damage in animals.[13]
- Carbon nanotubes (CNTs) have potential uses in a wide range of applications (e.g., materials, batteries, memory devices, electronic displays, transparent conductors, sensors, medical imaging). However, some scientists have expressed concerns that some CNTs exhibit properties similar to asbestos fibers, and might become lodged in organs (e.g., lungs), harming humans and animals.[14]

EHS CONCERNS ABOUT CARBON NANOTUBES AND OTHER FULLERENES

Much of the public dialogue about potential risks associated with nanotechnology has focused on carbon nanotubes (CNTs) and other fullerenes (molecules formed entirely of carbon atoms in the form of a hollow sphere, ellipsoid, or tube) since they are currently being manufactured and are among the most promising nanomaterials. These concerns have been amplified by some research on the effects of CNTs on animals and on animal and human cells. For example, researchers have reported that carbon nanotubes inserted into the trachea of mice can cause lung tissue damage;[a] that buckyballs (spherical fullerenes) caused brain damage in fish;[b] and that buckyballs can accumulate within cells and potentially cause DNA damage.[c]

There are scientists who have argued that experiments indicating CNT/fullerene toxicity are not conclusive. They suggest that toxicity reported by researchers may have resulted from uncharacterized contaminants in the samples resulting from the synthesis, purification, and post-processing methods used in the manufacture of CNTs. Thus, they assert, the experiments could be measuring the toxicity of non-nanoscale materials and, therefore, unfairly indicting nanoscale materials. They also contend that such non-nanoscale contaminants, if identified as toxic, potentially could be eliminated or controlled in the manufacturing process. The issue of contaminants is often cited by advocates for improved standards, reference materials, sensors, instrumentation, and other technologies for the characterization of nanoscale materials.

Some experiments have produced results that indicate CNTs/fullerenes are non-toxic. Research on single-walled carbon nanotubes (SWCNTs) by the Institute of Toxicology and Genetics in Karlsruhe, Germany, reported that, in three of four different types of tests conducted, SWCNTs did not show toxicity. In the fourth test, which appeared to indicate SWCNT toxicity, the researchers concluded that the results were a "false positive" and explained how the SWCNTs interacted with the materials in the assay to produce a misleading result. These researchers concluded that this result points to the need for careful selection of assays and the need for the establishment of standards for toxicity testing of CNTs and other nanomaterials.[d]

Work at Rice University's Center for Biological and Environmental Nanotechnology conducted in 2005 found cell toxicity of CNTs to be low, and that it could be reduced further through simple chemical changes to the surface.[e] Earlier research demonstrated that similar surface modifications of buckyballs reduced their toxicity. Nanotechnology may offer the potential to engineer around known and potential hazards by changing the size, molecular construction, or other property of a nanoscale material to make it safe or less hazardous. Experts advise that the potential to do so will require a thorough understanding of the properties of the various nanoparticles and their effects on humans and other organisms.

[a] Lam, C.W., James, J.T., McCluskey, R., and Hunter, R.L. "Pulmonary toxicity of single-wall carbon nanotubes in mice 7 and 90 days after intratracheal instillation," September 2003. http://www.ncbi.nlm.nih.gov/sites entrez?cmd=Retrieve&db=PubMed&list_uids=14514958

[b] Oberdörster, Eva. "Manufactured Nanomaterials (Fullerenes, C60) Induce Oxidative Stress in the Brain of Juvenile Largemouth Bass," April 2004. http://www.pubmedcentral.nih.gov/articlerender.fcgi?tool=pmcentrez&artid=1247377

[c] Magrez, A., Kasas, S., Salicio, V., Pasquier, N., Seo, J.W., Celio, M., Catsicas, S., Schwaller, B., and Forro, L. "Cellular Toxicity of Carbon-Based Nanomaterials," Nano Letters, 6(6):1121-1125, American Chemical Society, May 2006. http://pubs.acs.org/cgi-bin/abstract.cgi/nalefd/2006/6/i06/abs/nl060162e.html

[d] Wörle-Knirsch, J.M., Pulskamp, K., and Krug, H. F. "Oops They Did It Again! Carbon Nanotubes Hoax Scientists in Viability Assays," American Chemical Society, *Nano Letters,* Vol.6, April 2006.

[e] "Modifications render carbon nanotubes nontoxic," press release, Rice University, October 2005.

Importance of Addressing EHS Issues

Nanotechnology covers a wide swath of scientific fields, engineering disciplines, and technological applications. Sufficient knowledge has been developed about the useful properties of certain nanomaterials, how they can be manufactured, and how they can be applied in useful ways to enable commercial product development. In other areas of nanotechnology, fundamental research on nanoscale phenomena and processes is under way that may lead to greater understanding and beneficial applications in the years ahead. In general, however, nanotechnology is still an emerging field and there is a dearth of information about how nanoscale particles and devices might adversely affect human health, safety, and the environment. Accordingly, there is widespread agreement on the need for more research to better understand such implications.

In reviews of the NNI ,[15] both the National Research Council and the President's Council of Advisors on Science and Technology (PCAST) concluded that assessment of potential nanotechnology EHS risks is not possible due to the absence of information and tools. According to the NRC,

> it is not yet possible to make a rigorous assessment of the level of risk posed by [engineered nanomaterials]. Further risk assessment protocols have to be developed, and more research is required to enable assessment of potential EHS risks from nanomaterials.[16]

Similarly, PCAST concluded that

> it is premature to rigorously assess the levels of risk posed by engineered nanomaterials. Adequate tools are being developed but are not yet in place.[17]

Leaders of the NNI have argued strongly that to achieve the economic, societal, and EHS benefits of nanotechnology the nation must concurrently address its potential adverse effects. According to then-Under Secretary of Commerce for Technology Phillip J. Bond, a leading Administration advocate for the NNI,

> Addressing societal and ethical issues is the right thing to do and the necessary thing to do. It is the right thing to do because as ethically responsible leaders we must ensure that technology advances human well-being and does not detract from it. It is the necessary thing to do because it is essential for speeding technology adoption, broadening the economic and societal benefits, and accelerating and increasing our return on investment.[18]

This is a view shared by many in the business community. A 2006 survey of business leaders in the field of nanotechnology indicated that nearly two-thirds believe that "the risks to the public, the workforce, and the environment due to exposure to nano particles are 'not known,'" and 97% believe that it is very important or somewhat important for the government to address potential health effects and environmental risks associated with nano technology.[19]

The Project on Emerging Nanotechnologies (PEN) has warned that bad practices in nanotechnology research or production may result in a nanotechnology accident that would

> chill investment, galvanize public opposition, and generally lead to a lot of hand wringing on the part of governments who are betting large sums of money on the nanotech revolution.[20]

Successfully addressing EHS issues is seen as vital for those potentially exposed to nanoscale materials (e.g., consumers, researchers, manufacturing workers, the general public), businesses, and investors for a variety of reasons:

- protecting and improving human health, safety, and the environment;
- enabling accurate and efficient risk assessments, risk management, and cost- benefit trade-offs;
- ensuring public confidence in the safety of nanotechnology research, engineering, manufacturing, and use;
- preventing a problem in one application area of nanotechnology from having negative consequences for the use of nanotechnology in unrelated application areas due to public fears, legislative interventions, or an overly-broad regulatory response; and
- ensuring that society can enjoy the widespread economic and societal benefits that nanotechnology is believed by many to offer.

In addition, the U.S. regulatory environment for nanotechnology could be an enabler for innovation and contribute to a strong, sustainable economy by creating predictability, accurately assessing risks and benefits, and fostering the swift movement of safe products into the market. Such an environment is likely to favor nanotechnology-related investments and innovative activities in the United States by domestic and foreign stakeholders, as opposed to nations where such regulatory conditions do not exist.

Conversely, if the U.S. regulatory environment is not handled effectively (i.e., if it lacks predictability, if regulatory approaches do not accurately assess risks and benefits, or if approval processes are too long or expensive) it could prove a major impediment to innovation, economic growth, and job creation, as well as posing a potential threat to health, safety, and the environment. In such a regulatory environment, investment capital may be driven away from nanotechnology, potentially beneficial products may not be developed, safe products may be denied regulatory approval, or unsafe products may be allowed to enter the market.

Alternatively, nanotechnology investments, research, and production may be driven to other nations with preferable regulatory environments. On the one hand, such a regulatory system might be more desirable to investors and companies because it is more predictable, more efficient, and less costly. In such a case, the United States might miss out on nanotechnology's potential economic benefits. On the other hand, if other nations' regulatory systems are more attractive to investors and producers because those systems under-regulate or do not regulate at all, then nanotechnology research, development, and production could present increased EHS risks worldwide.

SELECTED ISSUES FOR CONSIDERATION

Given the widespread agreement that nanotechnology EHS concerns must be addressed, discourse on how best to do so has focused on three main issues:

- federal investment in EHS research;
- federal regulation; and
- international engagement.

These issues are closely interrelated. For example, reliable EHS research is required by regulatory bodies to determine whether and how to regulate nanotechnology products. Since all nations face the same fundamental health, safety, and environmental issues, international coordination on EHS research could help accelerate development of a common body of knowledge through the sharing of results and reduction in redundant research. This shared knowledge could, in turn, inform regulatory decision making and perhaps improve the consistency of regulations among nations. Regulations, standards, and enforcement might need to be coordinated worldwide to protect workers and consumers as intermediate and final products are frequently produced along global supply chains and sold in industrial and commercial markets around the world. In addition, one nation's policies governing nanotechnology production, use, and disposal may have implications for nearby nations and, perhaps, for all nations.

Federal Investment in EHS Research

Current Funding Level

There is not a single, centralized source of EHS research funds that is allocated to individual agencies. Agency nanotechnology budgets are developed internally as part of each agency's overall budget development process. These budgets are subjected to review, revision, and approval by the Office of Management and Budget (OMB) and become part of the President's annual budget submission to Congress. The NNI budget—and the EHS component—is then calculated by aggregating the nanotechnology components of the appropriations provided by Congress to each federal agency. While there is some coordination of EHS-research budget requests through the Nanotechnology Environmental and Health Implications (NEHI) working group[21] and in OMB's budget development process, the decision process that establishes overall funding for nanotechnology EHS research is highly decentralized.

In FY2008, NNI funding for EHS implications research[22] was $58.6 million, approximately 3.9% of the total NNI budget of $1.49 billion. This represented an increase over the FY2007 EHS research level of $48.3 million (3.4% of the total NNI budget), and the FY2006 level of $37.7 million (2.8%), both in dollars and in share of total NNI funding. President Bush requested $76.4 million (5.0%) for EHS research in FY2009. NNI EHS research funding for FY2006 through FY2008, and the request for FY2009, is provided in **Table 1**.

NNI officials assert that the initiative also conducts EHS research as a part of its other research activities, but that these EHS investments are not easily quantified and thus are not reflected in the NNI's reported figure for EHS funding. PCAST agreed with this assertion in its 2008 assessment, arguing that

> In many instances, nanotechnology EHS research cannot be separated from the particular application(s) research and from the context for which a specific nanomaterial is intended. Such division is unproductive and neglects the whole benefit of research. Consequently, [PCAST] expects that a substantial fraction of nanotechnology research related to EHS will continue to take place under the auspices of agencies that fund applications R&D and may not be uniquely or exclusively identified as nanotechnology EHS research.... Furthermore, detailed reporting on the degree of relevance to EHS of such research is not necessarily critical to (and may actual hinder) overall prioritization and coordination.[23]

Table 1. NNI Environmental, Health, and Safety Research Funding, FY2006-2008, FY2009 Request

	EHS research, in current dollars	EHS research's share of total NNI budget
FY2006 (actual)	$ 37.7 million	2.8%
FY2007 (actual)	48.3 million	3.4%
FY2008 (estimated)	58.6 million	3.9%
FY2009	76.4 million	5.0%

Sources: "The National Nanotechnology Initiative: Research and Development Leading to a Revolution in Technology and Industry, Supplement to the President's FY2008 Budget," NSET Subcommittee, NSTC, OSTP, The White House, July 2007; "National Nanotechnology Initiative: FY2009 Budget and Highlights," NSET Subcommittee, NSTC, The White House, February 2008. http://www.nano.gov/NNI_FY09_budget_summary.pdf

Table 2. FY2006 NNI Funding for EHS Research by Research Needs Categories

Category	Estimated Funding
Instrumentation, Metrology, and Analytical Methods	$27 million
Nanomaterials and Human Health	$24 million
Nanomaterials and the Environment	$13 million
Health and Environmental Exposure Assessment	$ 1 million
Risk Management Methods	$ 3 million
TOTAL	**$67 million**

Source: Teague, E. Clayton, director, National Nanotechnology Coordination Office. Testimony before the Subcommittee on Research and Science Education, Committee on Science and Technology, U.S. House of Representatives. Hearing on "Research on Environmental and Safety Impacts of Nanotechnology: Current Status of Planning and Implementation under the National Nanotechnology Initiative." 110th Cong., 1st Sess., October 31, 2007.

Note: Numbers may not add due to rounding.

In 2007, OMB issued a one-time request to all NNI research agencies to report funding data on research related to the five categories identified in the NSET document, *Prioritization of Environmental, Health, and Safety Research Needs for Engineered Nanoscale Materials.*[24] Totals for EHS implications research spending identified in each of the five categories is shown below in **Table 2**. Preliminary analysis of this data by the NEHI working group indicated that $67 million was spent on EHS research in FY2006, in contrast to the reported figure of $37.7.

Critics (as detailed in the following section) assert that the current level of federal nanotechnology EHS research is too low and represents too small a share of the overall NNI budget. These critics argue that the current allocation of NNI funding may produce a flood of products for which there is inadequate information to assess and manage their EHS risks.

However, executive branch officials stress that the United States leads the world in EHS funding and, by inference, that the current funding level is adequate. White House Office of Science and Technology Policy (OSTP) director John Marburger asserted that the United States

> leads the world not only in spending for nanotechnology development, but also, by an even larger margin, in its investment in research to understand the potential health and safety issues.[25]

Similarly, NNCO director E. Clayton Teague asserted U.S. leadership in nanotechnology EHS research:

> During fiscal years 2005 through 2008, it is estimated that NNI agencies will have invested nearly $180 million in research whose primary purpose is to address the EHS implications of nanomaterials. With these investments, the United States leads all other countries by a wide margin in support of such research.[26]

Dr. Teague maintains that EHS research has been a top priority of the Administration and the NNI, citing, as an example, the annual R&D budget guidance memorandum sent by the directors of OMB and OSTP to departments and agencies. This memorandum identifies Administration priorities and is intended to help guide agency budget development for the following fiscal year. The OMB/OSTP memorandum to guide FY2006 agency budget development stated that

> In order to ensure that nanotechnology research leads to the responsible development of beneficial applications, agencies also should support research on the various societal implications of the nascent technology. In particular, agencies should place a high priority on research on human health and environmental issues related to nanotechnology and develop, where applicable, cross-agency approaches to the funding and execution of this research.[27]

The OMB/OSTP memorandum has included similar language in each succeeding year.

In their reviews of the NNI, both the NRC and PCAST concluded that federal EHS research funding should be expanded. According to the NRC assessment,

> To help ensure the responsible development of nanotechnology ... research on the environmental, health, and safety effects of nanotechnology [should] be expanded.[28]

PCAST acknowledged potential EHS risks in its first review of the NNI but found the federal government was "directing appropriate attention" and "adequate resources" to EHS research. In its second assessment, PCAST termed the current federal investment level in EHS "appropriate," but added that

> expanded EHS research, broad-based protocol development, and particularly standardization are necessary.... the funding level for EHS [should] continue to grow consistent with the needs identified in the NNI research strategy for nanotechnology EHS as well as the available capacity for quality research.[29]

Alternative Approaches

Various alternatives have been suggested for addressing the perceived shortcoming in EHS funding. One recommendation is requiring a fixed percentage of the NNI's total funding be devoted to EHS research. A figure of 10% has been proposed for this purpose by organizations such as the NanoBusiness Alliance and the Project on Emerging Nanotechnologies. If this proposal had been in effect in FY2008, the NNI would have been required to spend $149 million on EHS research, more than twice as much as the NSET-reported level of $58.6 million. In testimony before the House Committee on Science and Technology, Sean Murdock, executive director of the NanoBusiness Alliance, agreed with the level of funding represented by the 10% figure but argued the need for cross-agency flexibility in achieving it:

> The NanoBusiness Alliance believes that environmental, health, and safety research should be fully funded and based on a clear, carefully-constructed research strategy. While we believe that 10 percent of the total funding for nanotechnology research and development is a reasonable

> estimate of the resources that will be required to execute the strategic plan, we also believe that actual resource levels should be driven by the strategic plan as they will vary significantly across agencies.[30]

Others have suggested a different approach, proposing fixed dollar amounts or minimum levels. For example, the Environmental Defense Fund has called for $100 million or more in federal nanotechnology EHS research funding.[31]

In its 2008 assessment, PCAST disagreed with both approaches:

> growing research in nanotechnology EHS must be strategic, guided by ... a comprehensive set of scientifically determined priorities and needs rather than arbitrary percentages or funding figures.[32]

By establishing a 10 percent requirement (or setting a figure of $100 million for total EHS funding), the United States could greatly accelerate the growth in EHS research spending. In testimony before Congress in 2007, PCAST co-chair Floyd Kvamme warned against such a rapid increase:

> In general, increasing funding too rapidly does not lead to equivalent increases in high quality research. It is crucial to note that EHS research also depends on advances in non-EHS areas, such as instrumentation development and basic research on nanomaterials.[33]

Some non-governmental organizations (NGOs) have advocated for a more restrained approach to nanotechnology research and development. They assert that the federal government is pushing ahead too quickly in developing nanotechnology and encouraging its commercialization and use without sufficient knowledge and understanding of EHS implications and how they might be mitigated.[34] They argue that the very characteristics that make nanotechnology promising also present significant potential risks to human health and safety and the environment. Some of these groups argue for application of the "precautionary principle,"[35] which holds that regulatory action may be required to control potentially hazardous substances even before a causal link has been established by scientific evidence.[36] In 2006, Friends of the Earth warned that

> The early warning signs surrounding nanotoxicity are serious and warrant a precautionary approach to the commercialization of all products containing nanomaterials ... . there should be a moratorium on the further

> commercial release of sunscreens, cosmetics and personal care products that contain engineered nanomaterials, and the withdrawal of such products currently on the market, until adequate public, peer-reviewed safety studies have been completed, and adequate regulations have been put in place ... [37]

The Action Group on Erosion, Technology, and Concentration (ETC Group) has called for a moratorium on the conduct of nanotechnology R&D and use of commercial products incorporating man-made nanoparticles:

> Given the concerns raised over nanoparticle contamination in living organisms, Heads of State ... should declare an immediate moratorium on commercial production of new nanomaterials and launch a transparent global process for evaluating the socio-economic, health and environmental implications of the technology.[38]

In 2003, the ETC Group expanded the breadth of its proposed moratorium:

> In the absence of toxicology studies, ETC Group believes that governments must also urgently consider extending the moratorium to products that place consumers in direct contact with synthetic nanoparticles through their skin, lungs or digestive systems.[39]

In contrast to these views, a report prepared by the NSET Subcommittee concluded that conducting EHS research in parallel with the development of nanomaterials and their applications will help to ensure the full, safe, and responsible realization of the promise of nanotechnology.[40]

In 2003, then-Under Secretary of Commerce for Technology Phillip J. Bond addressed calls for a moratorium or slowdown in nanotechnology R&D, casting the issue in ethical terms:

> Those who would have us stop in our tracks argue that it is the only ethical choice. I disagree. In fact, I believe a halt, or even a slowdown, would be the most unethical of choices.... Given the promise of nanotechnology, how can our attempt to harness its power at the earliest opportunity—to alleviate so many of our earthly ills—be anything other than ethical? Conversely, how can a choice not to attempt to harness its power be anything other than unethical?[41]

Management of Federal EHS Research

In order to manage the Federal EHS portfolio, policymakers will need to establish research priorities. In this regard, the NRC recommended that

> Assessing the effects of engineered nanomaterials on public health and the environment requires that the research conducted be well defined and reproducible and that effective methods be developed and applied to (1) estimate the exposure of humans, wildlife, and other ecological receptors to source material; (2) assess effects on human health and ecosystems of both occupational and environmental exposure; and (3) characterize, assess, and manage the risks associated with exposure.[42]

In 2005, PCAST concluded that EHS research should give highest priority to workplace exposure. PCAST noted

> the greatest likelihood of exposure to nanomaterials is during manufacture, and therefore [we] agree with the prioritization of research on potential hazards from workplace exposure.[43]

Several years later, in its 2008 assessment, PCAST reiterated this point stating, "the greatest risk of exposure to nanomaterials at present is to workers who manufacture or handle such material," but also acknowledged a broader range of risks:

> environmental, health, and safety risks in a wide range of settings must be identified and the necessary research performed so that real risks can be appropriately addressed.[44]

Some stakeholders have asserted that a comprehensive approach to federal EHS research has been hampered by the lack of an NNI roadmap for these efforts.[45] In general, these stakeholders seek a multi-year roadmap with specific milestones, metrics, and funding levels. Such a roadmap, they assert, would contribute to a more coordinated approach among agencies and between the executive branch and Congress on the magnitude, timing, prioritization, and management of federal EHS research.

NNI officials argue that the NSET Subcommittee, the coordinating body for the NNI, has developed an EHS research strategy and articulated it in three reports (see text box, "NNI EHS-focused Reports"), though they acknowledge that these documents do not constitute a roadmap. At an October 2007 hearing of the House Subcommittee on Research and Education,[46] some Members of

Congress expressed concerns about the time required by the National Nanotechnology Coordination Office to produce a prioritized, detailed implementation plan for NNI EHS research. While acknowledging the challenges faced by the NNCO in developing consensus among the 25 NNI agencies, some Members suggested that these challenges were emblematic of the need for a more top-down approach to EHS research.

Opposition to an EHS roadmap stems primarily from doubts of the practicality and efficacy of such an approach. Some argue that it is unlikely that OMB would commit to a multi-year, multi-agency roadmap accompanied by specific funding levels. Such an approach would depart from the current executive branch annual budget development process and reduce OMB's flexibility in future years. In addition, agencies often have to respond to new requirements based on emergent circumstances, Congressional direction, or other factors. Agency funding is often redirected from planned efforts to new, often imminent, priorities. The need for such redirection of funding could impede the achievement of roadmap milestones and metrics or, conversely, impede the movement of funding to new priorities.

NNI EHS-FOCUSED REPORTS

Environmental Health and Safety Research Needs for Engineered Nanoscale Materials, published in September 2006, identified the research and information needed to enable sound risk assessment and risk management decision making with respect to nanoscale materials and products that incorporate them.

Prioritization of Environmental, Health, and Safety Research Needs for Engineered Nanoscale Materials, published in August 2007, identified five broad categories of EHS research and information needs, and five specific research areas in each category.

The National Nanotechnology Initiative: Strategy for Nanotechnology-related Environmental, Health, and Safety Research, published in February 2008, defined the NNI's strategy for addressing priority research on EHS aspects of nanomaterials. The document reviewed current agency research using the taxonomy developed in the second report; identified research gaps; and articulated a framework for prioritizing research, implementing the strategy, and coordinating agency efforts.

To overcome the obstacles associated with the development of a roadmap by the agencies, some have suggested the National Academies produce such a roadmap. Some assert that this approach worked well with respect to the development of a federal research roadmap to reduce EHS uncertainties associated with airborne particulate matter. Others argue that the particulate matter effort focused only a narrow field and covered research conducted by only a single agency (EPA); in contrast, nanotechnology spans a broad range of materials and applications across many fields, and requires EHS research efforts by several agencies.

In February 2007, 19 environmental and business organizations, large and small companies, and research organizations signed a letter to the Senate Appropriations Subcommittee on Interior, Environment, and Related Agencies requesting $1 million be appropriated for the development of a federal roadmap and research strategy. The letter recommended that this work be done by the National Institute of Environmental Health Sciences (NIEHS).[47]

The Senate Appropriations Committee report (S.Rept. 110-91) accompanying the Department of the Interior, Environment, and Related Agencies Appropriations Act, 2008[48] urged the Environmental Protection Agency (EPA) to

> contract or enter into a cooperative agreement with the National Academy of Sciences' Board on Environmental Studies and Toxicology within 90 days of enactment to develop and monitor implementation of a comprehensive, prioritized research roadmap for all Federal agencies on environmental, health and safety issues for nanotechnology.[49]

A Cooperative Approach to Addressing EHS Concerns

Some organizations have taken a cooperative approach to promote EHS research. For example, the Environmental Defense Fund, an environmental advocacy group, partnered with the American Chemistry Council, a trade group, to issue a Joint Statement of Principles in June 2005 that recognizes the "significant societal and sustainable development benefits" expected from nanotechnology, while calling for a multi-stakeholder dialogue to achieve the timely development of nanomaterials "in a way that minimizes potential risks to human health and the environment." The statement also called for increased federal investments in EHS research and development of an international effort to standardize testing protocols, hazard and exposure assessment approaches, and nomenclature and terminology ... to maximize resources and minimize inconsistent regulation of nanomaterials.[a]

There is general agreement among stakeholders that these activities can contribute to creating an environment where research results can be reliably shared and compared, to protecting human health and safety, and to creating a common language about nanotechnology that increases clarity in the sharing of ideas and information. However international standardization efforts are often time- and resource-consuming, and can divert resources from more pressing needs. In addition, such efforts can be used by nations and other organizations for competitive advantage (e.g., by securing the adoption of a favorable standard, slowing others' progress).

In June 2007, the Environmental Defense Fund and DuPont issued a Nano Risk Framework "to assist with the responsible development and use of nanotechnology and to help inform global dialogue on its potential risks."[b] The framework is a six-step process to identify, address, and manage potential risks: (1) describe the material and the intended application; (2) profile the material's lifecycle in the application; (3) evaluate associated risks; (4) assess risk management options; (5) decide on and document actions; and, (6) regularly review new information and adapt actions accordingly.[c]

[a] *Environmental Defense and American Chemistry Council Nanotechnology Panel: Joint Statement of Principles*, Comments on EPA's Notice of a Public Meeting on Nanoscale Materials, June 23, 2005.

[b] "DuPont and Environmental Defense Launch Comprehensive Tool for Evaluating and Addressing Potential Risks of Nanoscale Materials," press release, E. I. du Pont de Nemours and Company, June 21, 2007. *http://vocuspr.vocus.com/VocusPR30/Newsroom/*Query.aspx?SiteName=DupontNew&Entity=PRAsset&SF_PRAsset_PRAssetID_EQ=106677&XSL=PressRelease&Cache=False

[c] *Nanorisk Framework*, Environmental Defense-DuPont Nano Partnership, June 2007.

The process used to develop research priorities and the federal EHS budget has also raised management concerns. As discussed earlier, the federal nanotechnology EHS research portfolio results from research funding requests made by individual agencies pursuing their missions and by decisions made in the Congressional appropriations process. Informal research coordination among EHS funding agencies occurs through the NEHI working group and more formally through the OMB budget development process. Some proponents for an integrated federal EHS research effort have called for a more top-down approach. The Woodrow Wilson Center's Project on Emerging Nanotechnologies (PEN) has been a leading advocate on this issue. PEN's chief science advisor, Andrew Maynard, asserted that

> to realize nanotechnology's benefits ... the federal government needs a master plan for identifying and reducing potential risks. This plan should include a top-down risk research strategy, dedicated and sufficient funding to do the job, and the mechanisms to ensure that resources are used effectively.[50]

PEN has recommended increasing the authorities of the NEHI working group to empower it to develop and implement the top-down research plan, a minimum of $100 million over two years to fund the research, and a full-time director to support the NEHI working group.

Responding to the PEN recommendation, E. Clayton Teague, director of the NNCO, testified before Congress that there was a consensus among NNI agencies that a centralized office with budgetary authority to oversee the NNI's EHS research program would have significant detrimental effects. According to Dr. Teague,

> No one agency or centralized organization would have the breadth of scientific expertise and knowledge of regulatory authorities and needs currently represented by the 20 agencies participating in the NEHI working group.
>
> Creation of a new central authority would undermine the existing successful interagency coordination.
>
> Moving the management of all nanotechnology EHS research into a single office would likely decouple such research from related efforts within NNI agencies and from the knowledge base in the agencies that is currently networked into the NNI's EHS research effort.
>
> Creating a separate office would, on the one hand, give mission agencies a disincentive for doing nanotechnology-related EHS research. They would reasonably assume that another agency is responsible, and they therefore could redirect their limited resources to address other priorities. A likely result could be that the level of research would actually decrease. Conversely, creating a separate office could lead to duplicative work being funded, thereby wasting tax dollars and not optimizing progress.[51]

Dr. Maynard counters that "it should be possible to develop a functional structure that enables agencies to work within a broader plan." According to Maynard, while a centralized office is not necessary,

> top-down leadership with authority and the ability to ensure resources get to where they are needed is necessary.... [Such] leadership does not take away from agencies' expertise and missions, but rather empowers agencies to do the best they can, while coordinating and partnering as effectively as possible with each other.[52]

PROJECT ON EMERGING NANOTECHNOLOGIES RECOMMENDATIONS

The Project on Emerging Nanotechnologies (PEN), a joint venture of the congressionally-chartered Woodrow Wilson Center for International Scholars and the Pew Charitable Trusts, has produced inventories of both nanotechnology-based products and government-funded EHS research. PEN has asserted the need for more EHS research, more aggressive oversight, and a more centralized federal government approach to funding EHS research.

In addition, PEN contends that the increasing complexity of systems incorporating nanoparticles with multiple functions will make the behaviors more complex and difficult to predict. To minimize the likelihood of a nanotechnology accident, PEN made the following recommendations:

- Creating a Nano Safety Reporting System where people working with nanotechnology can anonymously report safety issues and concerns. PEN states that the information gleaned from this system could be used to inform the design of educational materials, better structure technical assistance programs, and provide an early indicator of emerging safety issues.
- Creating technologies that provide an early-warning system to allow for risk to be assessed early in research efforts. Such a technology might enable low-cost, fast-screening for novel properties that would allow for risk assessment integrated and concurrent with the R&D process.

- Pushing information out to small businesses, start-ups, and laboratories that, due to their size and resources, are unlikely to be able to devote significant resources to EHS issues. PEN states that existing assistance programs could be used to deliver this information, as well as the development of peer-to-peer mentoring programs within industrial supply chains.
- Application of lessons learned in other technology areas to make nanotechnology more inherently safe, using strategies such as multiple levels of protection, learning from failures, not oversimplifying the complex, awareness of operations, and building in resilience to prevent cascading of errors.

Source: Rejeski, David, director, Project on Emerging Nanotechnologies. "Nanotech Safety 101 or How to Avoid the Next Little Accident," paper, Workshop on Disaster Prevention, Harvard University, April 27, 2006.

Federal Regulation

Some have raised concerns about whether current laws, regulations, and authorities are adequate to protect human health, safety, and the environment from potential adverse implications of nanotechnology. Several factors may affect the ability of the regulatory system to keep pace with advances in technology, both broadly and specifically with respect to nanotechnology.

Broadly, market forces have increased the pace of global innovation, challenging institutions' ability to identify and cope with the societal implications of rapid change. Speed-to-market has become a driving factor in competition for many industries as a result of the entry of new and nimble competitors in the global marketplace, increased public and private investments in R&D, global models of innovation, increased flows of scientific and technical knowledge, and greater numbers of scientists and engineers around the world. In addition, growing global markets enable companies to recoup their investments faster and enable earlier investments in subsequent generations of technology, further accelerating the pace of innovation. The increased pace, scope, and complexity of technological innovation may pose challenges to the existing regulatory system. While these factors may affect a broad range of technologies, nanotechnology may be especially affected due to the rapid growth in public and private R&D

investments in the field since the year 2000 and the potential for nanomaterials to be used in a wide array of products.

Nanotechnology also may pose unique challenges to the regulatory system. For example, historically, regulatory agencies have defined a chemical by its chemical composition, usually without regard to its particle size. In contrast, the essence of nanotechnology is that a material may exhibit different properties at the nanoscale than it does at a bulk, molecular, or atomic scale. (See text box, "Unique Properties Emerge at the Nanoscale.") Accordingly, questions are being raised by representatives of the scientific, advocacy, and regulatory communities about how an EHS research portfolio might be structured when particle size may affect a material's properties, whether it may be necessary to incorporate particle size into regulatory regimes, and how this might be accomplished given the vast spectrum of particle sizes that might affect the characteristics of a particular material.

Some argue that EHS concerns about nanotechnology products can be handled under existing laws and regulations, while others see legal obstacles to adequate EHS regulation. In both of its assessments of the NNI, PCAST concluded that existing regulatory authorities were adequate for the current activities; that appropriate regulatory mechanisms should be used to address instances of harmful human or environmental effects of nanotechnology; and that new regulatory policies related to nanotechnology should be rational, science-based, and consistent across the federal government. Similarly, Sean Murdock, executive director of the NanoBusiness Alliance, asserted that

> The apparatus for effective nanotechnology regulation is largely in place through various statutes and agencies, but it lacks data and resources. To enable these agencies and for the nanotech regulation effort to succeed we must increase the level of funding available to them for nanotech environmental, health and safety research; coordinate efforts between agencies; establish metrics and standards that can be used to characterize nanomaterials; conduct ongoing research; and more.[53]

Others believe that new laws and regulations, or modifications to existing ones, may be required. J. Clarence Davies, senior advisor to the Project on Emerging Nanotechnologies and former EPA Assistant Administrator for Policy, Planning, and Evaluation argued that

> Nanotechnology is difficult to address using existing regulations. There are a number of existing laws—notably the Toxic Substances Control Act; the Occupational Safety and Health Act; the Food, Drug and Cosmetic Act;

and the major environmental laws (Clean Air Act, Clean Water Act, and Resource Conservation and Recovery Act)—that provide some legal basis for reviewing and regulating [nanotechnology] materials. However, all of these laws either suffer from major shortcomings of legal authority, or from a gross lack of resources, or both. They provide a very weak basis for identifying and protecting the public from potential risk, especially as nanotechnologies become more complex in structure and function and the applications become more diverse.

A new law may be required to manage potential risks of nanotechnology. The law would require manufacturers to submit a sustainability plan which would show that the product will not present an unacceptable risk.[54]

UNIQUE PROPERTIES EMERGE AT THE NANOSCALE

Scientists have discovered that elements and materials with the same chemistry can exhibit fundamentally different properties at the nanoscale. For example, platinum, which exhibits no magnetism in its bulk form, hows significant magnetic properties in nanoscale clusters of 13 atoms. The optical properties of gold also can change with particle size. At 10 nanometers, gold particles absorb green light and appear red, not gold.

Not only can nanoscale particles differ in properties from bulk material with the same chemical composition, they may also differ from other anoscale materials with the same chemical composition. For example, the melting point of an element—which was believed to be constant regardless of the element's particle size—can change with particle size. Nanotechnology research has demonstrated that the melting temperature of gold decreases when the particle's radius drops below 10 nanometers (from a melting temperature of approximately 1,000°C at 10 nanometers to approximately 500°C at 2 nanometers).

Source: Roduner, Emil. "Nanoscopic Materials: Size-Dependent Phenomena," University of Stuttgart, Germany, August 2006.

Davies further asserts that new mechanisms and institutional capabilities—including research programs, tax breaks, acquisition programs, and regulatory incentives—are needed to encourage beneficial applications of nanotechnology.

In developing the regulatory structure, some in the business and financial communities assert that stability and predictability are key characteristics for attracting investment and spurring commercial applications. According to Matthew Nordan, vice president of Lux Research, the

> ambiguity surrounding environmental, health, and safety regulation of nanoparticles is hampering commercialization. Firms do not want to play a game whose rules may change at any time.... That doesn't mean they want more regulations or more onerous regulations. They're just looking for a roadmap on how federal agencies such as the EPA or OSHA [Occupational Safety and Health Administration] plan to approach nanoparticles.[55]

Some tension exists between the goals of promoting the development of nanotechnology, ensuring the global competitive position of the United States, addressing potential EHS implications of nanotechnology, and coping with the unique challenges nanotechnology poses to the current regulatory regime. To prevent health and safety concerns from becoming an impediment to innovation, some suggest that health and safety research and regulation must be done near-concurrently with product development, keeping pace with the speed of innovation. Alternatively, others argue that the potential health, safety, and environmental implications are either unknown or of such significance that EHS research and regulation must precede nanotechnology development and commercialization. "By the time monitoring catches up to commerce the damage will already have been done," asserted Ian Illuminato, health and environment campaigner for Friends of the Earth.[56] AFL-CIO industrial hygienist Bill Kojola warned that

> Even though potential health hazards stemming from exposure have been clearly identified, there are no mandatory workplace measures that require exposures to be assessed, workers to be trained, or control measures to be implemented. [Nanotechnology] should not be rushed to market until these failings are corrected and workers assured of their safety.[57]

The National Research Council assessment of the NNI acknowledged the need for additional reproducible, well-characterized EHS data to inform risk-based guidelines and best practices and warned that until such information is available precautionary measures should be taken to protect the health and safety of workers, the public, and the environment.[58]

In its 2008 assessment of the NNI, PCAST asserted that risk research must not be considered in isolation, but rather in the context of the overall risks and benefits of a particular material or technology. This perspective is shared by many industry advocates who argue that regulatory decisions must balance the potential risks associated with a nanotechnology product against the benefits it delivers and the risk it displaces. Further, they maintain that nanotechnology products should not be held to a higher standard than non-nanotechnology products. PCAST also noted that manufacturers and sellers of nanotechnology products had responsibilities for ensuring workplace and product safety, and asserted that the NNI has a vital role in supporting federal regulatory agencies by providing them with EHS research results.

A description of selected nanotechnology EHS activities of federal regulatory agencies is provided in **Appendix A**.

International Engagement

International engagement on EHS issues is believed by many to be important to the responsible development and successful commercialization of nanotechnology. NNI officials assert that the United States has played a central role in convening international efforts to address EHS concerns. In its 2008 assessment, PCAST encouraged the NNI to coordinate its efforts with other nations to avoid duplication and to leverage investments, characterizing such work as "noncompetitive."[59]

Federal agencies have engaged internationally (e.g., with agencies of other nations, international organizations, standards organizations) across a wide range of nanotechnology-related areas, including standards, nomenclature, and EHS research. **Appendix B** provides an overview of selected international engagement efforts of NNI agencies related to environmental, health, and safety issues.

Advocates for international engagement assert a variety of potential benefits. For example, transparency and/or harmonization of standards and regulations may contribute to assurance of global supply chains and market confidence in nanotechnology products. Increased globalization of production and markets means that companies and consumers around the world are increasingly part of a common network. Manufacturers of final products generally rely on inputs from multiple suppliers in their global supply chains. The reliability of a final product often depends on the reliability of inputs, such as materials or components. Transparent and common standards and regulations may help to ensure the integrity of supply chains and final products. While this is an issue for a variety of non-nanotechnology products (e.g., the recent discovery of lead-tainted toys and other products imported from China), nanotechnology may present a unique challenge in that at least some nanoscale particles can be incorporated into materials and products in ways that cannot be easily detected or detected at all. Thus, producers and the consumers they serve must rely, in large measure, on standards and regulatory systems to ensure that nano scale materials are properly produced and represented throughout the supply chain. In the absence of such standards and regulatory systems, producers may not be able to rely on inputs or may incur additional costs for testing and verification; substandard inputs may be incorporated in final products making them underperform or unsafe, and possibly resulting in loss of market confidence and/or potential litigation; or nanotechnology materials may be incorporated without disclosure.

Internationally agreed upon standards could also contribute to greater comparability of research results, improving understanding of EHS-related aspects of nanotechnology, and promoting regulations that help protect human health and the environment. Common standards and nomenclature also may contribute to more effective global collaboration in nanoscale science, engineering, and technology R&D, accelerating the realization of nanotechnology's economic and societal potential.

Global engagement may help to establish a common environment for the development and production of nanotechnology products and to promote access to global markets. In the absence of such an environment, some nations may seek to attract investments in their markets by adopting lower environmental, health, and safety standards and regulations.

Finally, while much remains unknown about the transport and fate of nanoscale materials released into the environment, it is possible that countries and populations other than those where research and production activities take place may be affected. Efforts to promote the adoption of best practices in nanotechnology research, production, use, disposal, and recycling may protect human health and the environment worldwide.

International engagement on EHS research may pose problems, including the time, cost, difficulty, and alleged ineffectiveness of such collaborations. For example, while some advocates assert the need for swift action in advancing EHS research, international engagements often entail slow processes. Also, given the strong U.S. position in nanotechnology, broadly, and in nanotechnology EHS research, specifically, some may argue that other countries have little to contribute, that such efforts tax limited federal EHS financial and human resources, and that such diffusion of resources may slow overall EHS progress. Others might assert that international engagement efforts focused explicitly on nanotechnology are unnecessary given the wide variety of existing mechanisms and pathways for sharing academic research and environmental, health, and safety information across national borders.

Some may oppose international engagement efforts because they lack faith in the goodwill of some participating parties due to the potentially strong national interests at stake (e.g., military applications, economic growth, job creation). In 2003, then-Under Secretary of Commerce for Technology Phillip J. Bond questioned whether global calls for a slowdown in nanotechnology R&D to address environmental, health, and safety concerns are intended to allow other nations to close the nanotechnology leadership gap with the United States:

> I wonder very often if there are really calls for a slow-down so that other governments and countries might catch up.[60]

Others assert that the research required to understand and address EHS implications may be closely linked to applications-related R&D to create nanotechnology materials, products, or processes. In such cases, companies and countries may be reluctant to reveal EHS concerns and efforts, to cooperate in EHS research, or to share results as such actions may reveal competitive strategies, provide information others might use to compete against them (e.g., insights into promising materials or manufacturing processes), or result in unwanted scrutiny by regulators.

Concluding Observations

Advocates and critics agree that potential environmental, health, and safety implications of nanotechnology must be addressed if the full economic and societal benefits of nanotechnology are to be achieved. There is also general agreement that the current body of knowledge of how nanoscale materials might affect humans and the environment is insufficient to assess, address, and manage the potential risks. While there is agreement on the need for more EHS research, there are differing views on the level of funding required, how it should be managed, and related issues.

Congress is currently considering legislation, H.R. 554, that would reauthorize and amend the 21st Century Nanotechnology Research and Development Act, the appropriations bills that fund the NNI agencies' nanotechnology EHS research, and H.R. 820, the Nanotechnology Advancement and New Opportunities Act. Congress may use these opportunities to further address nanotechnology EHS implications issues, including: How much should the federal government appropriate for EHS research? How can the federal EHS research investment be better accounted for? How should the research be prioritized? Should the research be more centrally managed? How can EHS research results and best practices be shared more broadly? Can voluntary programs effectively provide needed information about industrial nanotechnology production activities? How can efforts to develop common nomenclature and standards be improved? Are existing laws, regulations, guidelines, and regulatory structures adequate? Is there sufficient coordination among federal regulatory agencies? What types of international engagement on nanotechnology research and regulatory issues could best foster responsible development of nanotechnology?

NANOTECHNOLOGY EHS-RELATED LEGISLATION IN THE 111TH CONGRESS

Two bills introduced in the 111th Congress contain provisions that seek to address nanotechnology EHS concerns. The following section summarizes selected EHS-related provisions of these bills.

H.R. 554—National Nanotechnology Initiative Amendments Act of2009

H.R. 554, the National Nanotechnology Initiative Amendments Act of 2008, was introduced on January 15, 2009, and referred to the House Committee on Science and Technology. This act would revise the 21st Century Nanotechnology Research and Development Act in a variety of ways, several of which specifically address nanotechnology EHS concerns. The legislation:

- directs the National Nanotechnology Coordination Office to develop and maintain a public database of NNI EHS projects, including the agency funding source and funding history;
- requires the National Nanotechnology Advisory Panel (NNAP) to be established as a "distinct entity" (the NNAP's functions are currently performed by the President's Council of Advisors on Science and Technology), and requires the establishment of a subpanel to assess

whether societal, ethical, legal, environmental, and workforce concerns are adequately addressed by the NNI;

- directs that the National Research Council, as part of its triennial review of the NNI, evaluate the adequacy of the NNI's efforts to address ethical, legal, environmental, human health, and other appropriate societal concerns;
- requires the designation of an associate director of the White House Office of Science and Technology Policy to serve as Coordinator for Societal Dimensions of Nanotechnology with responsibility for developing an annual research plan for federal nanotechnology EHS activities, monitoring and encouraging agency EHS efforts, and for encouraging agencies to engage in public-private partnerships to support EHS research;
- requires certain interdisciplinary research centers supported under the NNI to include EHS research to develop methods for developing environmentally benign nanoscale products and processes, to foster the transfer of research results to industry, and to provide interdisciplinary study programs to educate scientists and engineers in these methods;
- directs NNI agencies to support the activities of standards setting bodies involved in the development of standards for nanotechnology, including authorizing agency reimbursement of travel costs of scientists and engineers participating in these activities; and
- requires activities supported under the NNI's Education and Societal Dimensions program component area to include environmental, health, and safety education in its informal, pre-college, and undergraduate nanotechnology education efforts.

H.R. 820—Nanotechnology Advancement and New Opportunities Act

H.R. 820, the Nanotechnology Advancement and New Opportunities Act, was introduced on February 3, 2009. Among its provisions, the bill would require the NNCO to produce an annual research strategy that establishes priorities for the development and responsible stewardship of nanotechnology, as well as providing recommendations regarding the funding required to implement the strategy.

On February 3, 2009, H.R. 820 was referred to the House Science and Technology Committee; the House Ways and Means Committee; the House Energy and Commerce Committee; and the House Homeland Security Committee.

APPENDIX A. SELECTED NANOTECHNOLOGY EHS ACTIVITIES OF FEDERAL REGULATORY AGENCIES

Several federal regulatory agencies have begun to grapple with the EHS issues raised by nanotechnology in their spheres of responsibility. Some critics argue that there is a potential conflict of interest among some regulatory agencies that are, on the one hand, conducting and promoting nanotechnology research and that are, on the other hand, responsible for regulating nanotechnology applications. The following section provides an overview of selected EHSrelated nanotechnology activities of federal regulatory agencies.

Environmental Protection Agency

The Environmental Protection Agency (EPA) co-chairs the NEHI working group of the NSET, along with the National Institute for Occupational Safety and Health (NIOSH), a research institute within the Department of Health and Human Services. EPA, which has both a research function and a regulatory function, has asserted a need for more information to assess the potential EHS impacts of most engineered nanoscale materials. According to EPA, this information is needed

> ...to establish a sound scientific basis for assessing and managing unreasonable risks that may result from the introduction of nanoscale materials into the environment.[61]

EPA is supporting research on the toxicology, fate, transport, transformation, bioavailability, and exposure of humans and other species to nanomaterials to obtain information for use in risk assessment, a central aspect of EPA's mission.[62]

EPA reports it is working collaboratively with stakeholders both domestically and internationally to address industrial chemical nanoscale materials. (International EHS collaboration is discussed in **Appendix B**.) One example of EPA's domestic work is its effort to establish a Nanoscale Materials Stewardship Program (NMSP). The purpose of the NMSP is to engage industry in a process that will foster effective federal government decision-making through the sharing of otherwise proprietary information about the characteristics, development, and manufacture of nanoscale materials. As envisioned by EPA, the program is designed primarily to engage manufacturers of nanoscale materials that would be considered existing chemical substances under the Toxic Substances Control Act (TSCA), but also encourages the participation of individuals and organizations working at a variety of stages of product development. EPA says that NMSP is intended to help provide a firmer scientific foundation for regulatory decisions by encouraging the development of key scientific information and appropriate risk management practices for nanoscale chemical substances.

According to EPA, the data acquired through NMSP will be used to gain an understanding of which nanoscale materials are produced, in what quantities, how they are used, and the data that are available for such materials. EPA maintains that its scientists will use data collected through this program, where appropriate, to aid in determining how and whether certain nanoscale materials or categories of nanoscale materials may present risks to human health and the environment. EPA states that NMSP is also intended to assist in the identification and adoption of risk management practices in the development and commercialization of nanoscale materials, to encourage the development of test data needed to provide a firmer scientific foundation for future work and regulatory/policy decisions, and to promote responsible development.[63]

EPA solicited comments on the NMSP from stakeholders in a July 2007 Federal Register Notice.[64] The business community has been supportive of the use of voluntary programs to address EHS risks of nanotechnology. The NanoBusiness Alliance states in its EHS research policy statement that "EPA and NIOSH should receive adequate funding to develop and implement their voluntary programs."[65] Other organizations have expressed frustration with the speed at which EPA is moving to implement the NMSP. At an EPA public meeting held in August 2007, Richard Denison, senior scientist for the Environmental Defense Fund, testified that

> As a government response to addressing the possible downsides of the nanotechnology revolution, [the NMSP is] simply 'too little, too late.'[66]

The Project on Emerging Nanotechnologies' J. Clarence Davies testified at the same meeting that while NMSP is

> ... potentially a useful initiative ... The delay in starting the NMSP is discouraging. It gives a signal that there really is no urgency, that the agency is in no hurry to start the voluntary program, much less institute an adequate regulatory system.[67]

Some observers say that past experience with other voluntary environmental programs shows that such efforts can produce benefits for both industry and government. For industry, voluntary programs may provide an opportunity to provide input into the regulatory process, to delay costly and constraining mandatory regulations, and to improve corporate goodwill. For government, voluntary programs may increase access to real-world data and information, may reduce the cost of data creation and/or collection, provide insights into new problems and about emerging industries, and provide a mechanism to control pollutants that are currently unregulated and for which jurisdiction may be hard to obtain.[68] Others maintain that voluntary programs can be counterproductive if they delay implementation of an adequate oversight system.

Multiple statutes govern EPA's authority to regulate nanotechnology materials and devices, including the Clean Air Act (CAA, 42 U.S.C. 7401 et seq); Clean Water Act (CWA, codified generally as 33 U.S.C. §§1251-1387); Federal Insecticide, Fungicide, and Rodenticide Act (FIFRA, 7 U.S.C.136-136y); and Toxic Substances Control Act (15 U.S.C. 2601 et seq.).[69] Important issues have been raised about the application of EPA's authorities to regulate nanotechnology.

Key issues revolve around TSCA, which authorizes regulation of chemical commerce.[70] Under the provisions of TSCA, producers of a "new" material must provide EPA with a premanufacture notification (PMN). EPA then has 90 days to approve manufacture, to require information from manufacturers, or to restrict chemical use. Other TSCA provisions permit EPA regulation of existing chemicals already in commerce, but these rely on EPA fact-finding and rulemaking before EPA can require testing or restrict uses. Several NGOs have urged EPA to consider all nanoscale materials "new" regardless of whether the material is on the EPA inventory list in its bulk form.[71] However, some nanotechnology materials have the same chemical composition as materials that are already in commerce, raising the question of whether the nanotechnology materials are "new" and thus subject to PMN requirements. With respect to this issue, EPA stated that

> EPA is considering how best to evaluate and, where appropriate, manage the risks associated with engineered nanoscale materials (NMs).... Nanoscale materials are "chemical substances" as defined under TSCA and are subject to the law unless otherwise excluded. Thus premanufacture notifications (PMNs) are required under TSCA prior to manufacturing a NM "new" chemical substance. To assist potential submitters, EPA is developing a general approach to the TSCA inventory status of nanoscale substances in making the distinction between "new" and "existing" chemicals that are nanoscale materials. EPA is also developing an umbrella approach for evaluating both new and existing chemicals in NMs.[72]

When nanomaterials are intended to control pests, including microbes, FIFRA may offer EPA more authority to regulate nanotechnology than TSCA, according to Lynn Bergeson, chair of the American Bar Association's Section on Environment, Energy, and Resources:

> Under TSCA, once a substance is on the approved inventory list, any use is legitimate, but FIFRA is use-specific. The EPA always has the authority to assess the risk of pesticides, regardless of the use.[73]

Applicability of FIFRA to nanotechnology products was one aspect of a November 2006 EPA ruling that a device that "incorporates a substance intended to prevent, destroy or mitigate pests" is considered a pesticide and is required to be registered under FIFRA. While the ruling is not unique to nanomaterials, it came in the context of advertising claims for a washing machine containing nanoscale silver ions that kill microbes. EPA's ruling made this appliance the first nanotechnology product to be regulated under FIFRA. However, claims for the pesticidal effectiveness of the washing machine have been removed from advertisements, possibly limiting EPA's ability to regulate the device as a pesticide under FIFRA.

Food and Drug Administration

A variety of current and future products that incorporate nanotechnology fall, or may fall, under the regulatory auspices of the Food and Drug Administration (FDA), including cosmetics, medical devices, foods, drugs, biological products, and combination products.[74] FDA anticipates that many of the nanotechnology products that the agency is likely to regulate will be combination products, such as drug-device, drug-biological, or device-biological products. According to FDA, it regulates products based on their statutory classification rather than the technology they employ, thus the agency may not provide regulatory consideration to a nanotechnology product until well after its initial development.[75] Also, some critics maintain that FDA's limited regulatory authority over certain categories of products may limit its authority to regulate nanotechnology products.

With respect to the need for unique tests or requirements for regulating nanotechnology products, FDA states that its existing requirements may be adequate for most nanotechnology products it expects to regulate. FDA has asserted that nanotechnology products are in the same size-range as the cells and molecules its reviewers and scientists deal with every day. The agency says that every degradable medical device and injectable pharmaceutical generates particulates that pass through the nanoscale size range during the processes of their absorption and elimination by the body. According to FDA, it has no knowledge of reports of adverse reactions related to the "nano" size of resorbable drug or medical device products. New tests or other requirements may be needed, according to FDA, if new risks are identified arising from new materials or manufacturing techniques. FDA has established a Nanotechnology Interest Group (NTIG) comprised of representatives from each of its centers to facilitate the regulation of nanotechnology products.[76] Others, in particular consumer groups, counter that FDA's resources are insufficient to adequately address the safety of emerging technologies in general, and that the agency's regulatory approach, particularly for cosmetics, dietary supplements, and other products for which pre-market review is not required, would not detect any problems until such products had been in use.[77]

FDA does not provide grants for nanotechnology research but does conduct research in several of its centers to understand the characteristics of nanomaterials and nanotechnology processes. FDA is also collaborating with NIEHS on studies, as part of the interagency National Toxicology Program (NTP), examining the skin absorption and phototoxicity of nano-sized titanium dioxide and zinc oxide preparations used in sunscreens.

FDA says that there currently is no international regulation of nanoproducts or the underlying nanotechnology. FDA participates in multinational organizations where cooperative work on nanotechnology has been proposed, including the Organization for Economic Cooperation and Development (OECD), ASTM International, and the International Organization for Standardization (ISO). FDA plans to work with its foreign regulatory counterparts to share perspectives and information on regulation of nanotechnology.[78]

National Institute of Environmental Health Sciences/National Toxicology Program

While not a regulatory agency, NIEHS is conducting nanotechnology EHS research that will support the missions of regulatory agencies. In particular, NIEHS serves as home to the interagency National Toxicology Program. The NTP's mission is to coordinate toxicological testing programs, develop and validate improved testing methods, develop approaches and generate data to strengthen scientific knowledge about potentially hazardous substances, and communicate with stakeholders.[79] In 2006, the NTP established the Nanotechnology Safety Initiative (NSI), a broad-based research program to address potential human health hazards associated with the manufacture and use of nanoscale materials. The goal of this research program is to evaluate the toxicological properties of major nanoscale materials that represent a cross-section of composition, size, surface coatings, and physical and chemical properties, and to use these as model systems to investigate fundamental questions concerning whether nanoscale materials can interact with biological systems and how they might do so.[80]

According to NTP, the NSI is focused on three areas of research with respect to specific types or groups of nanoscale materials:

- non-medical, commercially relevant and available nanoscale materials to which humans are intentionally being exposed, such as cosmetics and sunscreens;
- nanoscale materials representing specific classes (e.g., fullerenes and metal oxides) so that information can be extrapolated to other members of those classes; and

- subsets of nanomaterials to test specific hypotheses about a key characteristic (such as size, composition, shape, or surface chemistry) that might be related to biological activity.

Current NSI research activities are focused on metal oxides, fluorescent crystalline semiconductors (also known as quantum dots), fullerenes, and carbon nanotubes.

NTP has also established a Nanotechnology Working Group (NWG) to serve as a technical advisory body to provide a structured and formal mechanism for bringing stakeholders together to learn about NTP nanotechnology research related to public health, address issues related to that research, and promote dissemination of those discussions to other federal agencies, nanotechnology stakeholders, and the public. Another function of the NWG is to provide a mechanism for the public and interested parties to provide advice to the NTP Board of Scientific Counselors.

Occupational Safety and Health Administration/National Institute for Occupational Safety and Health

The mission of the Occupational Safety and Health Administration (OSHA), an agency of the Department of Labor, is to ensure the safety and health of America's workers by setting and enforcing standards; providing training, outreach, and education; establishing partnerships; and encouraging continual improvement in workplace safety and health. OSHA has not yet taken any regulatory actions with respect to nanotechnology.

The National Institute for Occupational Safety and Health (NIOSH), a part of the Department of Health and Human Services, is the lead federal agency conducting research and providing guidance on the occupational safety and health implications and applications of nanotechnology. NIOSH co-chairs the NSET's NEHI working group. NIOSH is not a regulatory agency, but its work directly supports OSHA and other regulatory agencies.

NIOSH states that its nanotechnology efforts are building on its experience in defining the characteristics, properties, and effects of ultrafine particles—such as welding fumes and diesel particulates—as well as its experience in conducting advanced health effects laboratory studies and in fostering industrial hygiene policies and practices. NIOSH has developed interim guidelines for working with nanomaterials. The agency asserts that these guidelines are consistent with the best scientific knowledge of nanoparticle toxicity and control. NIOSH also maintains a Nanoparticle Information Library with information on the health and associated properties of nanomaterials as an online resource for occupational health professionals, industrial users, worker groups, and researchers.[81]

NIOSH and OSHA are considering risk management approaches that do not rely on traditional exposure- and time-limits. These new approaches seek to maximize flexibility for innovation while ensuring the health and safety of workers.[82]

Consumer Product Safety Commission

The Consumer Product Safety Commission (CPSC) is charged with protecting the public from unreasonable risks of serious injury or death from certain types of consumer products.[83] CPSC asserts that potential safety and health risks of nanomaterials can be assessed under existing CPSC statutes, regulations and guidelines. Since the Consumer Product Safety Act (15 U.S.C. 2051 et seq.) and the Federal Hazardous Substances Act (15 U.S.C. 1261 et seq.) do not require pre-market registration or approval of products, CPSC does not evaluate a product's risk to the public until it has been distributed in commerce.

In August 2005, CPSC commissioners approved a nanotechnology statement which notes that nanotechnology presents challenges that "may require unique exposure and risk assessment strategies." The CPSC statement identified regulatory challenges, including identification of the specific nanomaterial in a product; the need to characterize the materials to which a consumer is exposed during product use, including an assessment of the size distribution of the materials released; and the application of toxicological data of appropriate particle sizes to assess health risks. However, the CPSC takes the position that it is unable to make any general statements about potential consumer exposure to nanomaterials or the health effects that may result from exposure to nanomaterials during consumer use and disposal due to the wide variation in potential health effects and the dearth of exposure and toxicity data for specific nanomaterials.[84]

APPENDIX B. SELECTED INTERNATIONAL ENGAGEMENT EFFORTS OF NNI AGENCIES

Federal agencies have engaged internationally (e.g., with agencies of other nations, international organizations, standards organizations) on a host of nanotechnology-related issues, with a focus on EHS-related efforts such as scientific research, standards, nomenclature and terminology. The following section provides an overview of some of these activities.

In June 2004, the U.S. government initiated and hosted the first International Dialogue on Responsible Research and Development of Nanotechnology in Alexandria, Virginia. The meeting was attended by representatives from 25 countries and the European Union.

The following year the NSET Subcommittee established the Global Issues in Nanotechnology (GIN) working group. In addition to monitoring foreign nanotechnology programs and promoting U.S. commercial and trade interests in nanotechnology, GIN was chartered to broaden international collaboration on nanotechnology R&D, including research on safeguarding the environment and human health. GIN representatives participated in the second International Dialogue on Responsible Research and Development of Nanotechnology hosted by the European Community (EC) in Brussels in July 2005. These meetings focused on clarifying issues and concerns of scientists, engineers, and policymakers working in nanotechnology around the world.

GIN representatives have also participated in nanotechnology-related activities of the Organization for Economic Cooperation and Development (OECD). In June 2005, chemical experts from 30 OECD countries participated in the Joint Meeting of Chemicals Committee and Working Party on Chemicals, Pesticides, and Biotechnology. Participants agreed to launch an international effort to coordinate assessment procedures for chemicals manufactured with nanotechnologies, to work toward linking national databases on high production-volume chemicals, and to establish a harmonized template for reporting hazard data needed for the notification and registration of new and existing chemicals, biocides, and pesticides. In December 2005, EPA hosted and chaired a second meeting of this group in Washington, D.C., on the safety of manufactured nanomaterials.

In October 2005, the United States proposed the creation of a Working Party on Nanotechnology within the OECD's Committee for Scientific and Technological Policy. Established in March 2007, the objective of this working party is to promote international co-operation that facilitates research, development, and responsible commercialization of nanotechnology in member countries and in non-member economies.[85] EPA is also participating in the OECD's Working Party on Manufactured Nanomaterials which was established in September 2006 to facilitate international collaboration on EHS issues related to manufactured nanomaterials.[86]

Another focus of U.S. international cooperation efforts has been in the development of nanotechnology standards. In response to a request from the White House Office of Science and Technology Policy, the American National Standards Institute (ANSI) established the Nanotechnology Standards Panel (NSP) in June 2004 to facilitate and coordinate nanotechnology standards development in the United States, focusing its initial work on nomenclature and terminology.[87] Subsequently, the International Organization for Standardization (ISO) established the Nanotechnologies Technical Committee, a parallel organization to ANSI's NSP. E. Clayton Teague, director of the NNI NNCO, chairs the ANSI-accredited Technical Advisory Group (TAG) to the ISO and leads the U.S. delegation.

The United States was selected to lead the ISO Technical Committee's Working Group on Health, Safety, and Environmental Aspects of Nanotechnologies. The Working Group has forwarded the NIOSH document "Approaches to Safe Nanotechnology," incorporating additional input from five other countries, to the ISO Technical Committee on Nanotechnologies (ISO TC 229) for a full review. If approved by the ISO Technical Committee, the document (re-titled "Health and Safety Practices in Occupational Settings Relative to Nanotechnologies") will be issued as an international Publicly Available Specification, an informational document available to all countries. NSET reports that significant progress on nanotechnology terminology and nomenclature has also been made by the TC 229 working group.

Federal government funding also contributed to the establishment of the International Council on Nanotechnology (ICON), a non-profit organization. ICON was established as an affiliate program of the NSF-funded, Rice University-based Center for Biological and Environmental Nanotechnology (CBEN), and has received funding from the National Science Foundation, the National Institutes of Health, and other private and non-profit organizations. ICON characterizes its purpose as seeking to catalyze global activities that lead to sound and responsible nanotechnology risk assessment, management, and communications. ICON has held EHS workshops, produced EHS reports, developed an online database of scientific findings related to the benefits and risks of nanotechnology, and designed and executed a survey of corporate nanotechnology EHS practices. In March 2007, ICON and CBEN jointly launched *The Virtual Journal of Nanotechnology Environment, Health, and Safety*, which contains citations and links to articles on the EHS impacts of nanotechnology.

End Notes

1 Congress defined nanotechnology in the 21st Century Nanotechnology Research and Development Act (P.L. 108-153) as, "the science and technology that will enable one to understand, measure, manipulate, and manufacture at the atomic, molecular, and supramolecular levels, aimed at creating materials, devices, and systems with fundamentally new molecular organization, properties, and functions." ASTM International, one of the largest voluntary standards development organizations, has defined nanotechnology as, "A term referring to a wide range of technologies that measure, manipulate, or incorporate materials and/or features with at least one dimension between approximately 1 and 100 nanometers. Such applications exploit those properties, distinct from bulk or molecular systems, of nanoscale components." One nanometer is about the width of 10 hydrogen atoms placed side-by-side, or approximately 1/100,0000 of the thickness of a sheet of paper.

2 Nanotechnology EHS applications refers to the beneficial use of nanotechnology to improve health, safety and the environment; EHS implications refers to known and potential adverse effects of nanoscale materials on health, safety and the environment.

3 Project on Emerging Nanotechnologies. Figure as of April 2008.

4 *Profiting From International Nanotechnology*, Lux Research, December 2006.

5 "Genetically Modified Crops and Foods," Friends of the Earth, January 2003. http://www.foe.co.uk/resource/ briefings/gm_crops_food.pdf

6 "Fact Sheet for Nanotechnology under the Toxic Substances Control Act," Environmental Protection Agency. http://www.epa.gov/oppt/nano/nano-facts.htm

7 "Cancer Nanotech Plan Gets Nod of Approval," Science, Vol. 305, July 23, 2004. http://www.sciencemag.org/ content/vol305/issue5683/s-scope.dtl#305/5683/461c

8 *A Matter of Size: Triennial Review of the National Nanotechnology Initiative,* National Research Council, 2006. p. 148.

9 *Nanosilver: A Threat to Soil, Water and Human Health*? Friends of the Earth, March 2007. http://www.foe.org/pdf/ FoE_Australia_Nanosilver_report.pdf

10 "Blood-Brain Barrier Breached by New Therapeutic Strategy," press release, National Institutes of Health, June 2007. http://www3.niaid.nih.gov/news/newsreleases/2007/bloodbrainbarrier.htm

11 "Nanotechnology Risks: How Buckyballs Hurt Cells," Science Daily, May 27, 2008. http://www.sciencedaily.com/ releases/2008/05/080527091910.htm

12 *National Nanotechnology Initiative: Research and Development Supporting the Next Industrial Revolution, Supplement to the President's FY2004 Budget*, Nanoscale Science, Engineering and Technology Subcommittee, Committee on Technology, National Science and Technology Council, The White House, October 2003. *http://www.nano.gov/ nni04_ budget_supplement.pdf*

13 Magrez, A., Kasas, S., Salicio, V., Pasquier, N., Seo, J.W., Celio, M., Catsicas, S., Schwaller, B., and Forro, L. "Cellular Toxicity of Carbon-Based Nanomaterials," *Nano Letters*, 6(6):1121-1125, American Chemical Society, May 2006. http://pubs.acs.org/cgi-bin/abstract.cgi/nalefd/2006/6/i06/abs/nl060162e.htmll

14 *Nanotechnology: The Future is Coming Sooner Thank You Think*, Joint Economic Committee, U.S. Congress, March 2007. p. 13. http://www.house.gov/jec/publications/110/nanotechnology

[15] The 21st Century Nanotechnology Research and Development Act (P.L. 108-153) requires a triennial assessment of the National Nanotechnology Program (in practice, of the NNI) by the NRC and a biennial assessment by PCAST, serving in its capacity as the National Nanotechnology Advisory Panel (NNAP). The act requires each assessment to include a review of the NNI's EHS activities. Three such assessments have been conducted, one by the NRC (*A Matter of Size: Triennial Review of the National Nanotechnology Initiative*, 2006) and two by PCAST (*The National Nanotechnology Initiative at Five Years: Assessments and Recommendations of the National Nanotechnology Advisory Panel*, May 2005; *The National Nanotechnology Initiative: Second Assessment and Recommendations of the National Nanotechnology Advisory Panel*, April 2008).

[16] *A Matter of Size: Triennial Review of the National Nanotechnology Initiative,* National Research Council, 2006. p. 90.

[17] *The National Nanotechnology Initiative: Second Assessment and Recommendations of the National Nanotechnology Advisory Panel*, President's Council of Advisors on Science and Technology, April 2008. p. 7.

[18] Bond, Phillip J., Under Secretary for Technology, U.S. Department of Commerce. "Preparing the Path for Nanotechnology: Addressing Legitimate Societal and Ethical Issues," keynote address, Nanoscale Science, Engineering, and Technology Subcommittee Workshop on Societal Implications of Nanoscience and Nanotechnology, December 3, 2003.

[19] "Survey of U.S. Nanotechnology Executives," Small Times Magazine and the Center for Economic and Civic Opinion at the University of Massachusetts-Lowell, Fall 2006.

[20] Rejeski, David, Director, Project on Emerging Nanotechnologies. "Nanotech Safety 101 or How to Avoid the Next Little Accident," paper, Workshop on Disaster Prevention, Harvard University, April 27, 2006.
http://www.nanotechproject.org/file_download

[21] NEHI is a working group of the Nanoscale Science, Engineering, and Technology (NSET) Subcommittee of the White House National Science and Technology Council (NSTC). The NSET Subcommittee is the coordinating body for the NNI. For additional information about the structure of the NNI, see CRS Report RL34401, *The National Nanotechnology Initiative: Overview, Reauthorization, and Appropriations Issues.*

[22] According to the NNCO, EHS research funding data included in Tables 1 and 2 of this report are for implications research only. The NNCO also states that the figures reported in **Table 1** may understate the NNI's EHS implications research by excluding funding for instrument research, metrology, and standards that support EHS implications research but are reported separately. (Source: Private communication between the NNCO and CRS.)

[23] *The National Nanotechnology Initiative: Second Assessment and Recommendations of the National Nanotechnology Advisory Panel*, President's Council of Advisors on Science and Technology, April 2008. p. 34.

[24] *Prioritization of Environmental, Health, and Safety Research Needs for Engineered Nanoscale Materials*, Nanoscale Science, Engineering, and Technology Subcommittee, Committee on Technology, National Science and Technology Council, The White House, August 2007.

[25] *Environmental, Health, and Safety Research Needs for Engineered Nanoscale Materials,* Nanoscale Science, Engineering, and Technology Subcommittee, Committee on Technology, National Science and Technology Council, The White House, September 2006. Cover letter.

[26] Teague, E. Clayton, director, National Nanotechnology Coordination Office. Testimony before the Subcommittee on Research and Science Education, Committee on Science and Technology, U.S. House of Representatives. Hearing on "Research on Environmental and Safety Impacts of Nanotechnology: Current Status of Planning and Implementation under the National Nanotechnology Initiative." 110th Cong., 1st Sess., October 31, 2007.

[27] "Updated Administration Research and Development Budget Priorities," memorandum, Office of Management and Budget and Office of Science and Technology Policy, The White House, August 12, 2004. http://www.whitehouse.gov/ omb/memoranda/fy04/m04-23.pdf

[28] *A Matter of Size: Triennial Review of the National Nanotechnology Initiative,* National Research Council, 2006. p. 92.

[29] *The National Nanotechnology Initiative: Second Assessment and Recommendations of the National Nanotechnology Advisory Panel*, President's Council of Advisors on Science and Technology, April 2008. pp. 7, 27.

[30] Murdock, Sean, executive director, NanoBusiness Alliance. Testimony before the Committee on Science and Technology, U.S. House of Representatives. Hearing on "The National Nanotechnology Initiative Amendments Act of 2008." 110th Cong., 2nd Sess., April 16, 2008.

[31] Denison, Richard A. "A Proposal to Increase Federal Funding of Nanotechnology Risk Research to at least $100 Million Annually," Environmental Defense, April 2005. http://www.edf.org/documents/4442_100milquestionl.pdf

[32] *The National Nanotechnology Initiative: Second Assessment and Recommendations of the National Nanotechnology Advisory Panel*, President's Council of Advisors on Science and Technology, April 2008. pp. 7, 27.

[33] Kvamme, Floyd, co-chair, President's Council of Advisors on Science and Technology. Testimony before the Subcommittee on Research and Science Education, Committee on Science and Technology, U.S. House of Representatives. Hearing on "Research on Environmental and Safety Impacts of Nanotechnology: Current Status of Planning and Implementation under the National Nanotechnology Initiative," 110th Cong., 1st Sess., October 31, 2007.

[34] Maynard, Andrew, chief science advisor, Project on Emerging Nanotechnologies, a joint venture of the congressionally-chartered Woodrow Wilson Center for International Scholars and the Pew Charitable Trusts. Testimony before the Subcommittee on Research and Science Education, Committee on Science and Technology, U.S. House of Representatives. Hearing on "Research on Environmental and Safety Impacts of Nanotechnology: Current Status of Planning and Implementation under the National Nanotechnology Initiative," 110th Cong., 1st Sess., October 31, 2007.

[35] The precautionary principle has been used in other countries on some issues and is the official policy in the European Union. For international agreements a precautionary approach is sometimes embraced. For example, the Biosafety Protocol to the 1992 Convention on Biological Diversity incorporates provisions applying the precautionary principle to the safe handling, transfer, and trade of genetically modified organisms. For further information, see CRS Report RL30594, *Biosafety Protocol for Genetically Modified Organisms: Overview*, by Alejandro E. Segarra and Susan R. Fletcher.

[36] "NGOs urge precautionary principle in use of nanomaterials," EurActiv.com, June 14, 2007, http://www.euractiv.com/en/environment nanomaterials/article-1 64619; Sass, Jennifer, "Nanotechnology and the Precautionary Principle," presentation, Natural Resources Defense Council, 2006. http://docs.nrdc.org/health

[37] *Nanomaterials, Sunscreens, and Cosmetics: Small Ingredients, Big Risks*, Friends of the Earth, May 2006. http://www.foe.org/camps/comm/nanotech/nanocosmetics.pdf

[38] "No Small Matter," Communique, ETC Group, May/June 2002. http://www.etcgroup.org/upload pdf_file/192 The ETC group is a non-governmental organization focused on the global societal impacts of emerging technologies.

[39] "No Small Matter II: The Case for a Global Moratorium," Occasional Paper Series, ETC Group, April 2003. http://www.etcgroup.org/upload

[40] *Environmental, Health, and Safety Research Needs for Engineered Nanoscale Materials,* Nanoscale Science, Engineering, and Technology Subcommittee, Committee on Technology, National Science and Technology Council, The White House, September 2006. p. vii.

[41] Bond, Phillip J., Under Secretary for Technology, U.S. Department of Commerce. "Nanotechnology: Economic Opportunities, Societal and Ethical Challenges," keynote address, NanoCommerce 2003, December 9, 2003.

42 *A Matter of Size: Triennial Review of the National Nanotechnology Initiative,* National Research Council, 2006. p. 92.

43 *The National Nanotechnology Initiative at Five Years: Assessments and Recommendations of the National Nanotechnology Advisory Panel*, President's Council of Advisors on Science and Technology, May 2005. p. 35.

44 *The National Nanotechnology Initiative: Second Assessment and Recommendations of the National Nanotechnology Advisory Panel*, President's Council of Advisors on Science and Technology, April 2008. p. 2.

45 Rejeski, David, director, Project on Emerging Nanotechnologies. Public comments on the Nanoscale Science, Engineering, and Technology Subcommittee's report, *Prioritization of Environmental, Health, and Safety Research Needs for Engineered Nanoscale Materials: An Interim Document for Public Comment*, September 12, 2007. *http://www.nanotech project.org/process/files/5891/nehi_comments_070912_final.pdf*

46 Hearing on "Research on Environmental and Safety Impacts of Nanotechnology: Current Status of Planning and Implementation under the National Nanotechnology Initiative." 110th Cong., 1st Sess., October 31, 2007.

47 An electronic copy of this letter, dated February 22, 2007, was provided to the Congressional Research Service (CRS) by the American Chemistry Council.

48 Incorporated as division F of the Consolidated Appropriations Act, 2008 (P.L. 110-161).

49 S.Rept. 110-91, p. 54.

50 Maynard, Andrew, chief science advisor, Project on Emerging Nanotechnologies. "Public Meeting on Research Needs and Priorities Related to Environmental, Health, and Safety Aspects of Engineered Nanoscale Materials," comments, January 4, 2007. http://www.nano.gov/html/meetings/ehs/uploads/20070103 _1505 _Nanotechnology _Maynard _NNCO _Comments.pdf

51 Teague, E. Clayton, director, National Nanotechnology Coordination Office. Testimony before the Subcommittee on Research and Science Education, Committee on Science and Technology, U.S. House of Representatives. Hearing on "Research on Environmental and Safety Impacts of Nanotechnology: Current Status of Planning and Implementation under the National Nanotechnology Initiative." 110th Cong., 1st Sess., October 31, 2007.

52 E-mail communication, November 21, 2007.

53 "Nanotechnology Leaders to Converge in Washington, D.C., This Week for NanoBusiness Alliance Public Policy Tour," article, nanotechwire.com, February 16, 2006. *http://nanotech wire.com/news.asp?nid=2929*

54 Davies, J. Clarence. *Managing the Effects of Nanotechnology*, Project on Emerging Nanotechnologies, January 2006. p. 3. http://www.nanotechproject.org/process/assets

55 "U.S. Risks Losing Nano Lead," article, physorg.com, July 6, 2005. *http://www. physorg. com/news4963.html*

56 "International Coalition Calls for Oversight of Nanotechnology," press release, Friends of the Earth, July 31, 2007. http://action.foe.org/dia/organizationsORG/foe/pressRelease.jsp?press

57 Ibid.

58 *A Matter of Size: Triennial Review of the National Nanotechnology Initiative,* National Research Council, 2006. p.11.

59 *The National Nanotechnology Initiative: Second Assessment and Recommendations of the National Nanotechnology Advisory Panel*, President's Council of Advisors on Science and Technology, April 2008. p. 33.

60 *Regional, State, and Local Initiatives in Nanotechnology*, Nanoscale Science, Engineering, and Technology Subcommittee, Committee on Technology, National Science and Technology Council, The White House, 2005. p. 33.

61 "Fact Sheet for Nanotechnology under the Toxic Substances Control Act," Environmental Protection Agency. http://www.epa.gov/oppt/nano/nano-facts.htm

[62] "Exploratory Research: Nanotechnology Research Grants Investigating Fate, Transport, Transformation, and Exposure of Engineered Nanomaterials: A Joint Research Solicitation - EPA, NSF, & DOE," Environmental Protection Agency. *http://es.epa.gov/ncer/ rfa/2007 /2007_star_nanotech.html*

[63] EPA notes that the National Research Council described "responsible development" in its first triennial review of the NNI as "the balancing of efforts to maximize the technology's positive contributions and minimize its negative consequences. Thus, responsible development involves an examination both of applications and of potential implications. It implies a commitment to develop and use technology to help meet the most pressing human and societal needs, while making every reasonable effort to anticipate and mitigate adverse implications or unintended consequences."

[64] "Nanoscale Program Approach for Comment," Environmental Protection Agency. *http://www. epa.gov/oppt/nano/* nmspfr.htm

[65] "Nanotech Environmental, Health and Safety: Progress and Priorities," NanoBusiness Alliance. http://www.nanobusiness.org/ehspolicy.php

[66] "EPA's Actions on Health Risks of Nanomaterials Called 'Too Little, Too Late,'" press release, Environmental Defense Fund, August 2, 2007.

[67] Davies, J. Clarence, testimony, EPA Public Meeting on Nanoscale Materials Stewardship Program, August 2, 2007. http://www.nanotechproject.org/file_download

[68] Berger, Michael. "Implementing successful voluntary nanotechnology environmental programs appears to be a challenge," Nanowerk LLC, November 29, 2007. *http://www.nano werk.com/spotlight/spotid=3476.php*

[69] For additional information, see CRS Report RL3 0798, *Environmental Laws: Summaries of Major Statutes Administered by the Environmental Protection Agency (EPA)*, by Susan R. Fletcher et al.

[70] For more information about TSCA and nanotechnology, see CRS Report RL341 18, *The Toxic Substances Control Act (TSCA): Implementation and New Challenges*, by Linda-Jo Schierow.

[71] "The EPA's Toxic Substances Control Act: What you must know," *Small Times*, September/ October 2007.

[72] "New Nanotechnology Products," Environmental Protection Agency. http://www.epa.gov/oppt/ar/20052006/ managing/new _nano.htm

[73] "EPA Regulates Nano Product, Not Nano Industry," *Small Times*, January 2007.

[74] For additional information, see CRS Report RL34334, *The Food and Drug Administration: Budget and Statutory History, FY1980-FY2007*, coordinated by Judith A. Johnson.

[75] "FDA and Nanotechnology Products," Food and Drug Administration. *http://www.fda.gov/ nanotechnology*

[76] Ibid.

[77] See, for example, Michael Taylor, *Regulating the Products of Nanotechnology: Does FDA Have the Tools It Needs?* The Project on Emerging Nanotechnologies, October 2006, at http://www.nanotechproject.org/news/archive/ is_fda_nanotech-ready. The FDA Science Board, Subcommittee on Science and Technology, designated nanotechnology as one of eight emerging technologies that are most challenging for FDA. See FDA Science Board, Subcommittee on Science and Technology, *FDA Science and Mission at Risk*, November 2007, p. 4, at http://www.fda.gov/ohrms/dockets/ac/07/briefing/2007-4329b_02_01_FDA Report on Science and Technology.pdf. The FDA Science Board is the advisory board to the FDA Commissioner.

[78] *FDA and Nanotechnology Products: Frequently Asked Questions*, Food and Drug Administration, U.S. Department of Health and Human Services. *http://www.fda.gov/ nanotechnology*

[79] "Toxicology in the 21st Century: The Role of the National Toxicology Program," Update, National Toxicology Program, January 2004. http://ntp.niehs.nih.gov/ntp/htdocs/Liaison/2004JanLO_News.pdf

[80] *Toxicology in the 21st Century: The Role of the National Toxicology Program,* Department of Health and Human Services, February 2004. http://ntp.niehs.nih.gov/ntp/main_pages/NTPVision.pdf

[81] "Nanotechnology at NIOSH," National Institute for Occupational Safety and Health. http://www.cdc.gov/niosh/ topics/nanotech/

[82] *Environmental, Health, and Safety Research Needs for Engineered Nanoscale Materials,* Nanoscale Science, Engineering, and Technology Subcommittee, Committee on Technology, National Science and Technology Council, The White House, September 2006.

[83] CPSC's regulatory authorities are provided by the Consumer Product Safety Act; Federal Hazardous Substances Act of 1960, as amended by the Toy Safety Acts of 1969 and 1984 and the Child Protection Amendments of 1966; Poison Prevention Packaging Act of 1970; Flammable Fabrics Act of 1953; and Refrigerator Safety Act of 1956. For additional information, see CRS Report RS22821, *Consumer Product Safety Commission: Current Issues*, by Bruce K. Mulock.

[84] "CPSC Nanomaterial Statement," Consumer Product Safety Commission, August 2005. http://www.cpsc.gov/library/ cpscnanostatement.pdf

[85] "OECD Work on Nanotechnology," Organization for Economic Cooperation and Development. http://www.oecd.org/sti/nano

[86] "Fact Sheet for Nanotechnology Under the Toxic Substances Control Act," Environmental Protection Agency. http://www.epa.gov/oppt/nano/nano-facts.htm; "OECD Work on Nanotechnology," Organization for Economic Cooperation and Development. *http://www.oecd.org/sti/nano*

[87] "ANSI Establishes Nanotechnology Standards Panel," American National Standards Institute, August 5, 2004. http://www.ansi.org/news_publications/news_story.aspx?menuid=7&articleid=735

CHAPTER SOURCES

The following chapters have been previously published:

Chapter 1 – This is an edited, excerpted and augmented edition of a United States Environmental Protection Agency, Office of Research and Development publication, EPA/600/S-08/002, dated January 24, 2008.

Chapter 2 - This is an edited, excerpted and augmented edition of a United States Congressional Research Service publication, Report Order Code RL34614, dated February 9, 2009.

INDEX

C

G

H

I

T

U

V

W

Y

Z